CAMBRIDGE LIBRARY COLLECTION

Books of enduring scholarly value

Technology

The focus of this series is engineering, broadly construed. It covers technological innovation from a range of periods and cultures, but centres on the technological achievements of the industrial era in the West, particularly in the nineteenth century, as understood by their contemporaries. Infrastructure is one major focus, covering the building of railways and canals, bridges and tunnels, land drainage, the laying of submarine cables, and the construction of docks and lighthouses. Other key topics include developments in industrial and manufacturing fields such as mining technology, the production of iron and steel, the use of steam power, and chemical processes such as photography and textile dyes.

The Architect, Engineer, and Operative Builder's Constructive Manual

Little is known about Christopher Davy (*c.*1803–49), despite his regular contributions to architectural and engineering magazines in Britain and America. Describing himself as an 'architect and teacher of architecture', he also took an interest in steam engines and railway construction. In this work, published in 1839, and using information gathered from experiments by the Board of Ordnance, Davy begins by describing the characteristics of the geology of England and Wales, with regard to its suitability for obtaining building materials and laying strong foundations. He describes the means by which soil and rock samples may be taken, and gives details relating to the construction of the foundations of St Paul's Cathedral on the troublesome London clay. Later chapters discuss the practicalities of pile driving, the use of concrete, and the properties of limestone. Reflecting the progress of technical knowledge in the early nineteenth century, the work features several illustrations of contemporary apparatus.

Cambridge University Press has long been a pioneer in the reissuing of out-of-print titles from its own backlist, producing digital reprints of books that are still sought after by scholars and students but could not be reprinted economically using traditional technology. The Cambridge Library Collection extends this activity to a wider range of books which are still of importance to researchers and professionals, either for the source material they contain, or as landmarks in the history of their academic discipline.

Drawing from the world-renowned collections in the Cambridge University Library and other partner libraries, and guided by the advice of experts in each subject area, Cambridge University Press is using state-of-the-art scanning machines in its own Printing House to capture the content of each book selected for inclusion. The files are processed to give a consistently clear, crisp image, and the books finished to the high quality standard for which the Press is recognised around the world. The latest print-on-demand technology ensures that the books will remain available indefinitely, and that orders for single or multiple copies can quickly be supplied.

The Cambridge Library Collection brings back to life books of enduring scholarly value (including out-of-copyright works originally issued by other publishers) across a wide range of disciplines in the humanities and social sciences and in science and technology.

The Architect, Engineer, and Operative Builder's Constructive Manual

Or, A Practical and Scientific Treatise on the Construction of Artificial Foundations for Buildings, Railways, Etc.

Christopher Davy

CAMBRIDGE
UNIVERSITY PRESS

University Printing House, Cambridge, CB2 8BS, United Kingdom

Cambridge University Press is part of the University of Cambridge.
It furthers the University's mission by disseminating knowledge in the pursuit of education, learning and research at the highest international levels of excellence.

www.cambridge.org
Information on this title: www.cambridge.org/9781108070690

This edition first published 1839
This digitally printed version 2014

ISBN 978-1-108-07069-0 Paperback

The original edition of this book contains a number of colour plates, which have been reproduced in black and white. Colour versions of these images can be found online at www.cambridge.org/9781108070690

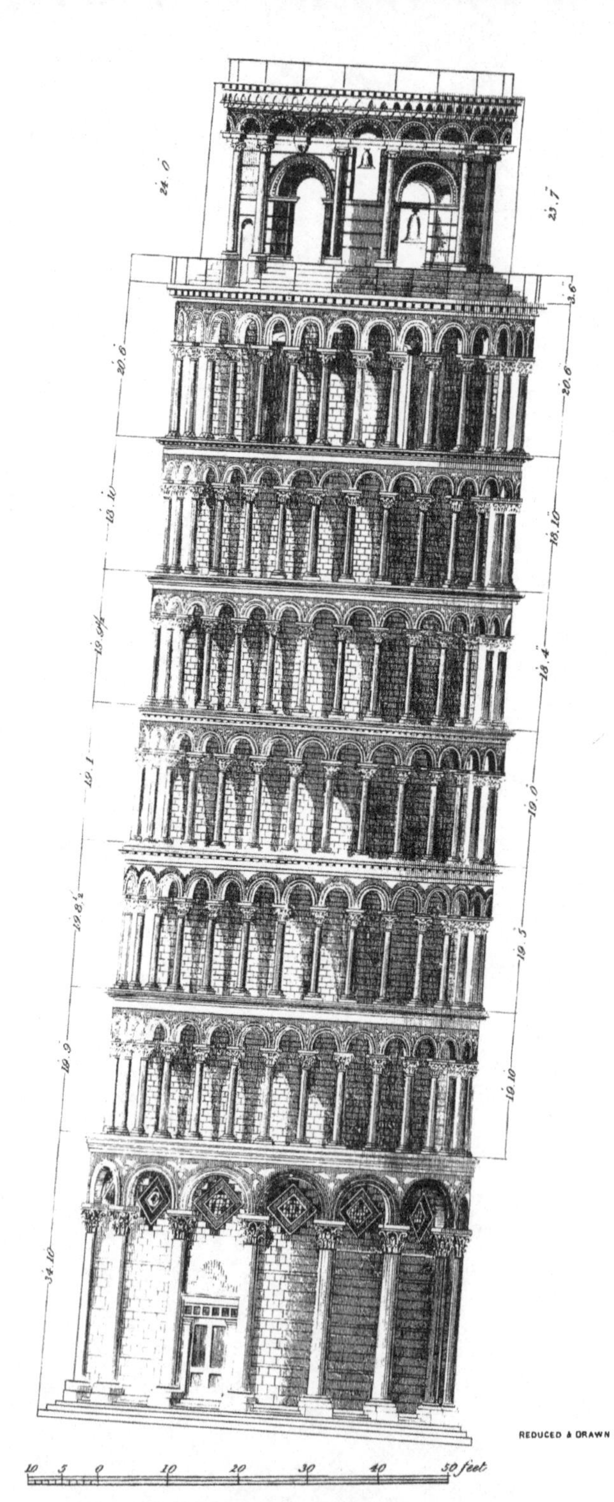

THE CAMPANILE AT PISA

Shewing the subsidence of the Foundation.

John Williams Library of Arts, Great Rufsell Street, Bloomsbury.

THE

ARCHITECT, ENGINEER,

AND OPERATIVE BUILDER'S

CONSTRUCTIVE MANUAL,

OR, A

PRACTICAL AND SCIENTIFIC TREATISE

ON THE

CONSTRUCTION OF ARTIFICIAL FOUNDATIONS FOR BUILDINGS, RAILWAYS, &c.

WITH A COMPARATIVE VIEW OF THE APPLICATION OF PILING AND CONCRETING TO SUCH PURPOSE; ALSO AN INVESTIGATION OF THE NATURE AND PROPERTIES OF THE MATERIALS EMPLOYED IN SECURING THE STABILITY OF BUILDINGS. THE WHOLE ILLUSTRATED BY EXAMPLES SELECTED FROM THE MOST IMPORTANT ARCHITECTURAL AND ENGINEERING WORKS OF THIS COUNTRY.

TO WHICH IS ADDED,

AN ANALYSIS OF THE PRINCIPAL LEGAL ENACTMENTS AFFECTING THE OPERATIONS OF THE PRACTICAL BUILDER. ILLUSTRATED BY NOTES OF CASES OCCURRING IN ACTUAL PRACTICE.

BY CHRISTOPHER DAVY,

ARCH. & C.E.

PART I.

LONDON:

JOHN WILLIAMS, LIBRARY OF FINE ARTS,

106, GREAT RUSSELL STREET, BLOOMSBURY.

1839.

DRURY, PRINTER,
Tooks Court, Chancery Lane, London.

CONTENTS OF PART I.

CONTENTS OF PART I.

[illegible]

ADVERTISEMENT.

The ramification of railways throughout the kingdom has so extended the sphere of the engineer's operations, as to render a knowledge of the geological formations peculiar to each county through which his passage lies, of considerable, nay, of essential importance. Geological knowledge is now to the engineer indispensable, inasmuch as it forms the source from which his supplies must be drawn, and, at the same time, forms the index by which may be ascertained the nature of that foundation upon which his construction is to rest.

In the ordinary routine of architectural practice, it is notorious that information of this kind has been rarely sought for until the time has arrived for carrying the intended works into execution. The uncertainty and delay incident to this mode of inquiry, has not unfrequently entailed ruinous consequences upon both architect and builder. If proofs were wanting, we should be prepared to show from specifications on the one, and breach of contracts on the other side, instances of defective information, and consequent difficulties and extravagance of execution, fully corroborative of this opinion. It is evident, therefore, that the engineer should be prepared with such information as will enable him to avail himself of the resources of each particular district, and to apply them with promptitude and decision.

The calcareous and siliceous deposits of this country have hitherto received but little attention from the profession; and we may add, that with respect to the nature and properties of the materials, we are as yet "upon the threshold of our knowledge."

The great and increasing demand for geological information applicable to architectural and engineering purposes, has induced the author to prefix to the following Treatise a Register* of the principal geological formations in each of the counties of England and Wales, with observations on the quality of the materials to be derived therefrom, and applicable to the purposes of building; also, the situation of the principal stone quarries, lime works, &c.

In compiling this "Register," the author has attempted to give a concise view of the resources of each county, in order that the draft of the specifications for important works may be made with

* In consequence of the extent occupied by this additional matter, (which has much exceeded our original plan,) the "Introduction," which is chiefly historical, is deferred till the publication of the second part of the "Treatise."

greater accuracy than heretofore, and that the builder may be directed to the sources from which he may obtain his materials with facility and economy.

To trace the extent and disposition of the strata, the engineer must necessarily consult an accurate geological map; for this purpose, the "Index Map," by J. Phillips, Esq., F.R.S., G.S., and the large map by Messrs. Walker, will supply much valuable information.

The reader will perceive, from the authorities which are given, that the author has, in the following compilation, freely availed himself of such information as was afforded by reference to the works of the most eminent writers on geological science; and having, also, visited and carefully examined most of the northern and western counties of England during the last two summers, he trusts that the variety of facts and data thus collected and brought under immediate observation, may meet with the approbation of the members of that profession to which he has the honour to belong.

3, *Furnival's Inn, Nov.*, 1838.

A REGISTER

OF THE PRINCIPAL GEOLOGICAL FORMATIONS IN EACH OF THE COUNTIES OF ENGLAND AND WALES, WITH OBSERVATIONS ON THE QUALITY OF THE MATERIALS TO BE DERIVED THEREFROM, AND APPLICABLE TO THE PURPOSES OF BUILDING; ALSO, THE SITUATION OF THE PRINCIPAL STONE QUARRIES, LIME WORKS, &c.

Northumberland.

THE county of Northumberland contains, 1. *Mountain or carboniferous limestone.* 2. *Coal.* 3. *Basaltic and porphyritic trap.* And, 4. *Old red sandstone,* with the accompanying beds of conglomerates, gritstones, variegated marls, and cornstones.

The building materials supplied by this county chiefly consist of limestone, limestone marble, and sandstones. The quarries are plentifully distributed throughout the county. Limestone abounds in the districts of Bamborough Ward and part of Glendale Ward, east of the river Till; it is also quarried at the following places:—Sprouston, Stodridge, Shilbottle, Long Franlington, Hartburn, Ryall, Corbridge, and Aldstone Moor. Blue, red, gray, and brown whinstone is procured from Bamboroughshire, the

Cheviot mountains, and adjacent places. It is generally employed for repairing the roads.

The limestones of Northumberland vary considerably in quality; they may be classified in the following order:—1. Highly crystalline. 2. Slightly crystalline. 3. Ferruginous. 4. Bituminous. And, 5. Calcareo-siliceous. The strongest lime is to be obtained from the calcination of Nos. 3 and 4. The crystalline varieties are best adapted for wrought masonry or architectural façades. For engineering works, or the substructure of important buildings, the sandstones, associated with the carboniferous deposits, are highly valuable. The following are the varieties:—1. *Slate sill*—a fine-grained micaceous, slaty rock, of a gray colour, used as a roofing slate in many villages of Northumberland and Durham. It is the uppermost bed in the section of Heley field. 2. *Freestone sills*—fine-grained quartzose sandstones, used for building. 3. *Hazles*—hard, ferruginous, fine-grained sandstones. 4. *Millstone grit*—a coarse, white, quartzose sandstone; it crops out on the Derwent, and is quarried for millstones. The quarries are at Muggleswick Fell, and between Wolsingham and Stanhope, in Weardale. A similar rock is found at Scramerstone, in the north-eastern part of Northumberland, and at Craster, near Howick. The castle of Dunstanborough is built with this stone. 5. *Grindstone sill*—a fine-grained yellowish sandstone found at Aldstone Moor (Durham), Coalcleugh, Allenheads (Northumberland), Nenthead, and on the summit of Cross Fell. The grindstones made from this material are said to be much inferior to those of Newcastle.

Below the limestone, in the Aldstone Moor section (Durham), are the following sandstones:—1. *Whetstone sill*—a fine-grained micaceous sandstone, at Burtreeford. 2. *Ironstone sill*—a ferruginous sandstone, containing an abundance of iron pyrites. 3. *Firestone*—a fine-grained, porous sandstone, used for furnaces. 4. *Pattison's sill*—a very hard, gray sandstone, containing mica. 5. The *coal sills*—similar to the above. And, 6. *Water sill*, or *tuft*—a very porous, soft, light-coloured sandstone.

The durability of sandstone chiefly depends upon its texture or state of induration. It has been remarked,* that when a porous sandstone is exposed to the weather, or placed in such situations where water may enter and penetrate its crevices, mere change of temperature will cause much mischief; but on the occurrence of a frost, the expansion becomes so great, that a single winter may cause a disaggregation of the entire mass.

Durham.

The county of Durham contains, 1. *Coal.* 2. *Millstone grit.* 3. *Magnesian limestone.* 4. *Mountain limestone.* And, 5. *Trap.*

* See Brande's Journal, vol. iii., p. 381.

The *great limestone* of Frosterley, in Weardale (Durham), is a dark gray shelly marble, extensively quarried for lime, ornamental purposes, &c.; it contains 96 per cent. of carbonate of lime. The *scar limestone* is similar to the above; it crops out in the Nent rivulet. The *cockle-shell limestone*—a dark, iron-gray, shelly limestone; it crops out on Aldstone Moor. The *Tyne-bottom limestone*—an encrinal limestone of three strata; it forms the bed of the Tyne for four miles, from Tyne Head to Garrigill Gate. *Robinson's great limestone* and the *Melmerby scar limestone*, are carboniferous beds of considerable thickness; the former is the lowest in the Dufton section, and the latter bassets out at Melmerby Cliff. At this place the bed is twenty-one fathoms in thickness. An account of the magnesian limestone of Durham, and an analysis of some specimens will be found in the 4th chapter of this Treatise. Magnesian limestone, when calcined, produces lime possessing peculiar properties, which are hereafter noticed. As a material for masonry, it is much superior to the tender oolitic varieties of the west of England. The atmosphere acts with much less intensity upon the fine-grained and magnesian limestone than upon the coarser and oolitic kinds. The superficial hardening which takes place upon the surface of the Bath oolite, when exposed to the weather, is very deceptive, inasmuch as it is followed by a rapid disaggregation of the stone. The *Castle* of *Clare*, in the county of Suffolk, is in part built with magnesian limestone. A portion of one of the bondstones from the *keep* of that structure, was procured by the author during the past summer. Upon examination it was found nearly perfect, while a specimen of oolite from the same building crumbled to calcareous sand. This building was erected some time previous to the year 1190.

The millstone grit and other sandstones of the county of Durham do not differ materially from the sandstones of Northumberland already described; such variations as occur being more interesting to the geologist than to the engineer, we omit any further notice of them.

The sandstone from the Heddon (Heddon-on-the-Wall) quarries, Northumberland, must not, however, be omitted in our Register. This stone is a hard, light-coloured quartzose sandstone, with rather a coarse grain, but admirably adapted for engineering works. Some of the earliest edifices in the county, constructed with this stone, yet remain to attest its durability.

The quarries recently opened at Heddon communicate with the Tyne river by a railway. The basalt, graywacke, new and old red sandstone, &c., are chiefly worked and employed for local purposes. The other materials will be described in the account of those counties where such formations assume a greater importance.

Cumberland.

Cumberland contains, 1. *New and old red sandstone.* 2. *Trap.* 3. *Granite.* And, 4. *Clay slate.* The principal free stone quarries (chiefly red and white siliceous and quartzose grits), are in the neighbourhood of Whitehaven; from whence it is shipped to various places. Quarries for grindstone grits are situated near Ivegill and Barngill. The blue slate quarries are at Bassenthwaite, Borrowdale, Buttermere, Cockermouth, and Ulpha. In this county there is also coal and its accompanying limestone, but the principal deposits from which building stone is drawn are the new and old red sandstone formations.

Westmoreland.

Westmoreland contains formations similar to the above, with the addition of a development of the Silurian system. These rocks furnish, 1. *Micaceous sandstones, limestones,* and *sandy shales.* 2. *Subcrystalline limestone shale.* 3. *Gritstones,* then *limestones* and *sandstones.* And, 4. *Calcareous* and *argillaceous flags.* The prevailing stratrum of the southern and eastern parts of the county is carboniferous limestone, associated with quartzose grits. Limestone marble is quarried within three miles of Kendal, also near Ambleside, and a bituminous limestone near Kirby Lonsdale and Kendal Fell. The great slate quarries occupy the western mountains; whinstone and slate is also procured from the north western mountains. The dark gray limestone of Windermere and its vicinity is quarried for lime—chimney pieces, &c., red felspathic granite, and another variety, much harder, and of a green colour, occurs at Wasdale.

Yorkshire.

Yorkshire contains, 1. *Coal.* 2. *Millstone grit.* 3. *Carboniferous limestone.* 4. *Trap.* 5. *Mountain* and *magnesian limestones.* And, 6. *Clays and crag.* Building materials may be obtained as follows. Quarries of freestone and millstone grit are to be found at and near Gatherley Moor, near Richmond, Renton near Boroughbridge, west of Sheffield, Penistone, Huddersfield, Bradford, Otley, Harrowgate, Ripley, Masham, Rainton quarry, Pateley quarry, Moorstone quarry, all within a few miles of Ripon. Quarries at Bramley, Mexborough, Holton, Kirby, Leeds, Idle, and Sheffield; all these are quartzose grits. Magnesian limestone quarries, Brodsworth, Quarry Moor, Ripon, and at Doncaster. Mountain limestone quarries—Roach Abbey and near Richmond, &c. Limestone and marble are also found in the Western Moorlands. A roofing flag slate (siliceous) is dug near Wensleydale. Lias is sufficiently near to be procured for lime.

Lancashire.

Lancashire contains, 1. *Coal.* 2. *Millstone grit.* 3. *Mountain* and *magnesian limestones.* 4. *New red sandstone.* The quarries of freestone are situated in that part of the county south of the sands. Flagstones are also raised in the same district. Blue slate quarries are most numerous in the northern parts of High Furness.

Anglesea.

Anglesea contains, 1. *Gneiss.* 2. *Mica schist.* 3. *Granite.* 4. *Coal.* 5. *Old red sandstone.* 6. *Carboniferous* or *mountain limestone.* 7. *Trap.* And, 8. *Clay slate.*

The quarries of the Isle of Anglesea have furnished a crystalline limestone, which appears well adapted for architectural façades; its use, however, has as yet been very limited, arising probably from the expense of carriage or freight. The most noted edifice erected with this material is the Birmingham Town Hall.

Caernarvon.

Caernarvon contains, 1. *Graywacke.* 2. *Trap.* 3. *Gneiss.* And, 4. *Mica schist.*

The great bulk of building materials employed in this country is derived from calcareous and siliceous deposits. The peculiar nature and composition of the primary rocks prohibit their employment for the ordinary purposes of building; in England, therefore, the supply drawn from these sources is almost exclusively confined to granite and slate. The county before us, and most others in the principality of Wales, exhibit these formations in considerable grandeur; we accordingly take this opportunity to offer a few general remarks on the structure and composition of the primary system of rocks, or at least such of them as appear to command the particular attention of the engineer. The primitive rocks are by geologists, usually classed as follows:—1. *Granite.* 2. *Gneiss.* 3. *Mica slate.* 4. *Clay slate.* 5. *Primitive limestone.* 6. *Primitive trap.* 7. *Serpentine.* 8. *Diallage rock.* 9. *Porphyry.* 10. *Quartz.**

The constituents of common granite will be found in the sixth chapter of this Treatise; on this subject it is only necessary to add, that the proportion per cent. of oxygen contained in this substance, and the particles of which it is composed, according to Mr. De la Beche, are as follows: 100 granite= 52 metallic bases + 48 oxygen. 100 quartz = 48.4 metallic + 51.6 oxygen. 100 felspar = 54 metallic bases + 46 oxygen. 100 mica = 56 me-

* "Jamieson's Mineralogy." For an account of the subdivisions of these rocks and other geological information concerning them, the reader is also referred to that work. "Phillip's Treatise on Geology," vol. i., and "De la Beche's Geological Manual."

tallic bases + 44 oxygen.* (See De la Beche's Geological Manual, 2nd edition.)

According to its locality, granite is known by various names; thus in Cornwall it is usually called "*moorstone*" (Smeaton); in Scotland, whinstone or sandstone (Jamieson), while among antiquarians the term granite has been applied to every rock possessing a granular structure. Popular opinion has conferred upon this material a character for durability exceeding every other kind of stone; it is nevertheless subject to rapid disaggregation and even decay, in proportion to the quantity and quality of its constituents. From numerous observations derived from chemical and mechanical experiments, it appears that the particles of granite most acted upon by the atmosphere, are those which form the great mass of its composition, namely, felspar.

In judging the quality of a *building stone* we may rest satisfied that the greater the quantity of argillaceous or earthy matter contained in it, the more certain and rapid will be its decomposition. Felspar appears to be a substance of an earthy or argillaceous character; thus, if we finely pulverize a portion of felspar, and add a small portion of water to the powder, we produce a paste forming a tolera-

* *Quartz* is a compound of a metallic basis, *silicium*, and the air or gas *oxygen*. *Felspar* is a compound of *silicium*, *calcium*, *potassium*, &c., each united with its own proportion of *oxygen*. *Mica* is a compound of *silicium*, *potassium*, *magnesium*, *calcium*, &c., similarly combined with *oxygen*. (Phillip's Treatise on Geology, vol. 1, p. 31.)

ble substitute for *clay*;* and it has been remarked,† that *clay slate* "is not very distinct, chemically speaking, from decomposed felspar which has lost or changed the condition of its *potash* by the operation of water." "Hence," continues Mr. Phillips, "under particular circumstances (which permit the access of alkali and great heat) that blue slate is actually transformed to white and glassy crystalline grains of *felspar*:" this discovery was made by Mr. W. V. Harcourt, of Yorkshire.

Although this granitic compound is, *per se*, liable to premature decay, it cannot be denied that granitic rocks offer greater resistance to the action of the atmosphere than most others. The difference of quality observable in granites is very striking; the coarse grained granites of Devon and Cornwall, compared with the finer grained and harder kinds of Aberdeen, exhibit a difference of strength of 14 to 22.‡

The strength of granite depends, first, upon the proportion of its constituent parts; and, secondly, upon the chemical state of those particles. The constituents of Herm granite (sienitic) are felspar, quartz, and hornblende, with a very small quantity of black mica; the whole being in a highly crystalline state. The Brittany granites generally contain a considerable quantity of mica, which from its fragile nature, or want of cohesion, deteriorates the qua-

* See "Phillip's Treatise on Geology," p. 124.

† Ibid.

‡ For a practical application of these varieties, the student is referred to Waterloo Bridge; the ballustrades are formed out of the Aberdeen granite.

lity of the stone; the Guernsey granites are tolerably free from this substance. A black granite, much used near London for mending the roads, is almost wholly composed of black masses of mica; in the sienite of Malvern, red felspar predominates, and in the Dublin granite, white opaque felspar, intermixed with a moderate quantity of black mica, are the chief constituents of the mass.

These varied proportions and qualities of the compounds create a very sensible difference in the quality of the stone; thus, in the sienitic granite of Herm the absence of micaceous particles gives place to hornblende, which is hard and crystallized in small prisms, (Lukis Trans. Inst. C. E.) thereby increasing the specific gravity of the stone. In the British granites, the felspar is mostly opaque; in the Herm granite it is crystallized and partly transparent; in the Dublin granite, the opaque felspar is the most prevalent; and in some of the English varieties exists in large cubical masses; again, in most British granites the quartz is comparatively small in quantity, but in the harder and more durable varieties, both quartz and felspar are abundant, and *transparent* or *crystallized.*

With these facts before us, it is not difficult to arrive at just conclusions respecting the quality of this material; but in the event of any doubts arising in the mind of the reader, we refer him to the 1st volume of the Transactions of the Institution of Civil Engineers, in which will be found a paper describing a series of important experiments on various kinds of granite, made under the direction of the late Mr. Bramah.

Among these experiments, we find that 2 cubes of Herm granite (4 × 4 × 4) weighing 6 lbs. 6 oz. were *fractured* with a mean pressure of 4.77 tons upon the superficial inch, and were *crushed* with a mean pressure of 6.64 tons per inch superficial, while the remaining specimens, consisting of Aberdeen (blue), Haytor, Dartmoor, Peterhead (red), and Peterhead (blue gray), declined gradually in strength, the two cubes of the last specimens weighing 5 lbs. 3½ oz., and 5 lbs. 4 oz., being fractured with a mean pressure of 2.86 tons upon the inch superficial, and crushed with a force of 4.36 tons upon the inch. For the locality of the principal English granites, the reader is referred to the account of the counties of Cornwall and Devon.

The constituents of mica schist and gneiss are analogous to those of granite, but arranged in laminæ; the former is of a slaty structure, composed of continuous sheets of mica, with grains of quartz : quartz rock and primary limestones are also interposed among the beds of mica schist (Phillips). Gneiss accompanies mica schist; its composition is very similar to granite, but the mica is disposed in patches, and occasionally much contorted, instead of being disseminated throughout the mass as in common granite.

These rocks are not generally suitable for building materials, and being sparingly exhibited in England, do not demand the attention of the engineer; we may, however, remark, that an assemblage of grains consisting of quartz, schist, felspar, and particles of mica, &c., agglutinated with a variable cement, form the "*psammites*" of Mr. Vicat. Those, he observes (p. 46, Treatise on Cements), are the most important which are slaty, of a yellow, red, or brown colour, fine grained, unctuous to the touch, producing a clayey paste with water. These belong to the primitive schistose formations, and cannot exist except in *situ*, being found in "beds or veins, forming part of the schist of which they are merely a decomposition."

In the composition of mortar, the schistose sands or "*psammites*" are feebly energetic,* namely, causing the mortar to *set* from the tenth to the twentieth day, and that acquires the hardness of dry soap after a year's immersion. Acid will act upon the *clay* of these sands, causing it to yield after maceration a portion of iron, and from $\frac{1}{10}$ to $\frac{3}{5}$ of its alumina.

The clay-slate and graywacke system exhibits a series of fine and coarse grained argillaceous, indurated, fissile rocks and sand, with conglomerates and limestone bands; the subdivisions consisting of dark flag-like slates, limestone, green slate fragmentary rocks, and dark soft slate. The transition limestone of Plymouth is associated with the graywacke series; it is a variegated limestone, having a splintery fracture, but like most calcareous rocks, is not proof against the action of water and

* Pure sands are *inert* substances. (See Vicat on Cements, by Smith.)

marine animals (vide the Plymouth breakwater); graywacke, basaltic trap, &c., are not to be recommended for engineering works, where other materials may be readily obtained.

Silurian rocks* (Murchison). The Ludlow rocks consist of beds of sandstone shale and limestone, with the following subdivisions:—1. *Laminated sandstone (Upper Ludlow rock).* 2. *Limestone (Aymestry limestone).* 3. *Sand, shale,* and *clay,* with concretions of *earthy limestone (Lower Ludlow rock).*

The Wenlock and Dudley rocks consist of strata of limestone and shale, with the following subdivisions. 1. A mass of highly concretionary *gray* and *blue sub-crystalline limestone (Wenlock* and *Dudley rocks).* 2. *Argillaceous shale, liver* and *dark gray coloured, rarely micaceous,* with *nodules of earthy limestone (Wenlock* and *Dudley shale).*

Horderly and Mayhill rocks. Conglomerates, sandstones and limestones. Subdivisions. 1. Thin bedded impure shelly limestone; and finely laminated, slightly micaceous greenish sandstone. 2. Thickly bedded, red, purple, green, and white freestones; conglomoritic quartzose grits, sandy and gritty limestones.

Builth and Llandeilo rocks. Beds of dark coloured flags, mostly calcareous, with some sandstone and schist. We now proceed to complete the

* This classification is from J. and C. Walker's elaborate and beautiful "Geological and Railway Map of England," &c., recently published.

enumeration of the Welsh counties, and the index to the foregoing classes of rocks and strata.

Denbigh.

The county of Denbigh contains, 1. *Clay-slate.* 2. *Graywacke.* 3. *Silurian rocks; and borders on the carboniferous limestone.*

Flint.

The county of Flint contains, 1. *Coal.* 2. *Millstone grit.* 3. *New red sandstone.* 4. *Silurian rocks.* 5. *Trap.* And, 6. *Carboniferous limestone* and *lias,* near the detached part of the county.

Merioneth.

The county of Merioneth contains, chiefly, *graywacke.*

Montgomery.

The county of Montgomery contains, 1. *Coal.* 2. *Trap.* 3. *Clay-slate* and *graywacke.* 4. *Silurian rocks.* And, 5. *Old red sandstone.*

Cardigan.

The county of Cardigan contains, 1. *Coal.* 2. *Millstone grit.* 3. *Silurian rocks.* And, 4. *New red sandstone.*

Radnor.

The county of Radnor contains, 1. *Trap.* 2. *Clay-slate.* 3. *Silurian rocks.* 4. *Graywacke.* And, 5. *Old red sandstone.*

Brecknockshire.

Brecknockshire contains, 1. *Old red sand stone.* 2. *Silurian rocks.* 3. *Clay-slate* and *graywacke.* 4. *Carboniferous limestone;* and skirts the coal of Glamorganshire.

Caermarthen.

The county of Caermarthen contains, 1. *Silurian rocks.* 2. *Clay-slate* and *graywacke.* 3. *Coal.* 4. *Old red sandstone.* And, 5. *Carboniferous limestone.*

Pembroke.

The county of Pembroke contains, 1. *Clay-slate* and *graywacke.* 2. *Coal.* 3. *Trap.* And, 4. *Old red sandstone.*

Glamorgan.

The county of Glamorgan contains, 1. *Coal.* 2. *Millstone grit.* 3. *Old red sandstone.* And, 4. *Lias.*

Monmouth.

The county of Monmouth contains, 1. *Silurian rocks.* 2. *Coal.* 3. *Old and new red sandstone.* And, 4. *Carboniferous limestone.*

For local purposes, the geological formations of the Welsh counties may supply abundant materials for engineering operations; but the peculiar composition of most of these deposits render them comparatively useless in any other district, particularly when it is remembered that great facilities are now afforded to the engineer in obtaining the more valuable and workable materials of the calcareous and carboniferous series.

Cheshire.

The county of Cheshire is mostly occupied by the new red sandstone formation. The sandstones of this formation are generally of a red colour, and not micaceous (Phillips). The subdivisions consist of marl, red sandstone, and conglomerates (Burr). The sandstones consist of a great diversity of colours, and, from this circumstance, the entire formation has received the name of "*variegated sandstone,*" "*bunter sandstein,*" (Werner); and "*red rock,*" "*red ground,*" and "*red marl,*" of other geologists. The new red sandstone is thus described by Mr. Aikin, in a paper on "The Wrekin and the Great Coal Field of Shropshire" (Geol. Trans., vol. i., p. 192):—"This rock consists, for the most part, of rather fine grains of quartz, with a few spangles of mica, ce-

mented by clay and oxyde of iron. Its colour i
generally brownish red, and it has but little cohe
sion, on which account, large tracts of loose, deep
sand are found in many parts of it. Sometimes it occurs nearly of a cream colour, and is then sufficiently hard to form an excellent building stone. It does not effervesce with acids, and, to the best of my knowledge, never contains shells or other organic remains."

The arenaceous deposits of the new red sandstone system exhibit considerable diversity of character; thus, at Kirkby-Stephen, the deposit contains a brecciated rock, with limestone fragments; at Nottingham, a pebbly conglomerate; at Runcorn, one with few pebbles: at Penrith Beacon, a coarse red grit; a fine kind, suitable for building, at Meriden-hill, near Coventry, and colour, white or green, (Warwickshire.)* Although the new red sandstone is not generally applicable for building, yet, in some districts, a very suitable and durable kind may be obtained. The principal quarries in Cheshire are at Runcorn, Manley, Peckforton, and adjacent places.

Buildings constructed with this material may be seen on the Liverpool and Manchester Railway; and a fine section of the rock, shewing its peculiar vertical fissures, may be examined with advantage at Olive Mount, the deep rocky cutting near the entrance of the Liverpool Tunnel. Stockport church, Cheshire, (Basevi, architect,) is mostly built with

* Phillips's "Treatise on Geology," p. 184.

Runcorn stone; and the bridge over the Dee, at Chester, by Harrison, is faced with a light-coloured stone from the quarries of Manley and Peckforten. The two courses of arch stones in the river abutments of Chester bridge are of Scotch granite (Craignair); the key course, with one on each side, and the quoins all through, are of Anglesea and similar limestone, from Wagbur quarries, at Burton, in Kendale. The tower of Kenton, in Devonshire, is also constructed with a durable stone belonging to the red sandstone series.* The unsightly and rusty appearance of some of the worst varieties of this stone, may be seen in many old buildings in the ancient city of Chester, including the cathedral.

The student should be informed that, in this country, the new red sandstone is always associated with the rock salt and gypseous deposits, hence the name, "saliferous." The limestones associated with the new red sandstone exhibit considerable variety; their colours are white, gray, smoky, yellow, and red (Phillips,) often containing much magnesia, hence their name, "magnesian limestones." The texture of these calcareous beds are oolitic, cellular, granular, powdery, and, in Nottinghamshire, granular and crystallized.

Derby.

Derbyshire contains, 1. *Coal*, consisting of alternations of *coal, sandstones, shales,* and *ironstone*

* Conybeare.

beds. 2. *Millstone grit,* namely, *quartzose grits* with *shales, coal, ironstone,* &c. 3. *Carboniferous* or *mountain limestone,* namely, *limestone* either in one mass or divided by *shales, gritstones, ironstones,* and *coal.* And, 4. *Trap.** A description of the carboniferous limestone will be found in Chapter 4 of this Treatise.

Mr. Farey, in his "Agricultural Survey of Derbyshire," enumerates forty-six limestone quarries in this county, and sixty-three lime kilns, from whence great quantities are conveyed and sold in this and some of the neighbouring counties. The largest quarries are at Ashover, Buxton, Crich, and Calver, near Barslow. A considerable quantity is sent from Calver into Yorkshire, and from Buxton into Cheshire and Staffordshire. The marble quarries are nineteen in number, and are situated near Bakewell and Matlock. The number of stone quarries are one hundred and thirty-eight, in some of which an excellent building stone may be obtained. The best quarries for this kind of stone are situated in the parish of Wingerworth. Millstone grit is obtained from nineteen quarries; and other finer kinds of sandstone from thirteen quarries. The best quarries for the latter are at Codnor Park and Woodthorpe.† The sandstones of the carboniferous series are amply described under the head "Northumberland."

* The magnesian limestone is adjacent to the county.

† Farey's "Survey of Derbyshire."

Nottingham.

The county of Nottingham contains, 1. *New red sandstone.* 2. *Magnesian limestone;* with *coal* on the Derby side and *lias* on the Lincoln side.

The prevailing deposit in this county is the new red sandstone, already described. Limestone is burnt in various parts of the county, both from the lias and magnesian deposits; from the former, on Beacon-hill, near Newark. The quarries for the best building and paving stone are situated at Mansfield, Maplebeck, Beacon-hill, and at Linby, a few miles to the south-west of Mansfield.* The new red sandstone of this county is mostly of a loose and friable texture.

Lincolnshire.

The county of Lincoln contains, 1. *Chalk.* 2. *Green sand.* 3. *Upper*, *middle*, and *lower oolite.* 4. *Lias;* with much *marsh* or *fenny* land. The materials peculiar to these formations will be found noticed under the several counties where the formations are more largely developed, namely, Surrey, Sussex, Kent, Northamptonshire, &c. It may be observed, however, that the lower chalk beds have a ferruginous or red stain.†

The strata immediately under the Lincolnshire

* "Agricultural Survey."

† A similar variety exists in the East Riding of Yorkshire.

chalk is described, by Mr. Bogg,* as consisting, first, of a coarse, ferruginous, pebbly, quartzose sand. Secondly, a stratum of oolitic limestone and calcareous clay, in nearly equal proportions. In certain parts of this bed, the clay divides the seams of stone into regular strata. The next stratum is that of a quartzose sandstone of different shades of colour. Notwithstanding these indications of strata capable of furnishing building materials, Lincolnshire does not possess any of sufficient importance to merit the attention of the engineer.

The soil of the Louth marshes consists of unstratified clay and sand, under which is the chalk; but a considerable denudation has been effected in Lincolnshire, whereby the upper oolitic beds have been removed. In consequence of the vast extent of fenny land in this county, the foundations of many of the oldest buildings rest upon piles; probably the earliest English examples of this mode of constructing foundations.

Rutland.

The county of Rutland contains, 1. *Lower oolite.* 2. *Lias.*

This county does not supply any building materials worth the particular attention of the engineer; but its situation with respect to other counties furnishing an ample supply of stone, lime, and brick, is such, that no difficulty is likely to arise

* Mr. Bogg, "On the Lincolnshire Wolds." Geol. Trans., vol. 3, p. 394.

from the want of materials in the immediate locality. The ferruginous sandy beds of the lower oolitic series supply calcareo-siliceous materials, which were formerly much employed in some of the adjoining counties. The materials derived from these beds are very inferior and unsightly for architectural purposes.* The sandstone series is generally separated from the great oolite by a thick clay (Conybeare).

Leicester.

The county of Leicester contains, 1. *Lias.* 2. *New red sandstone.* And, 3. *Sienitic granite, trap*, and *graywacke* of Mount Sorrel. Roofing slate is quarried at Swithland, east of Charnwood forest. The principal limestone quarries are at Bredon, Cloud-hill, and Barrow-upon-Soar. Freestone, of the new red sandstone species, is found in most parts of the county, also brick earth. Sienitic granite is quarried at Mount Sorrel; much of it is used for metalling the roads.

Stafford.

The county of Stafford chiefly contains *new red sandstone* and *coal.* The quarries situated in this county are very numerous. Excellent freestone, mostly derived from arenaceous deposits, is quarried largely. There are several quarries at

* The old buildings in and about Northampton, Banbury, &c., are built with stone procured from the strata alluded to.

Gornal, Bilston, Sedgley, Tixall, Wrottesley, Brewood Park, Pendeford, and the Rowley-hills. The stone from the latter place is much employed for paving; it is called Rowley rag-stone, and is found in heaps similar to the "stone shatter," or Kentish rag. In the neighbourhood of Tamworth there are several quarries, within a mile and a half of each other, from which is raised a durable building stone, provincially termed "whinstone." The railway viaduct now building at one end of Tamworth, is being constructed with this stone. The viaduct crosses the Tame river and two roads, and rises to an elevation of twenty feet above the town, and consists of eighteen square apertures of thirty feet span each, and a skew arch of sixty feet span. The stone is a grit of fine texture, and brownish colour. The quarries for this stone are at Dost-hill and Stokes quarry, Amington. Limestone abounds in this county. Extensive lime works are situated at Caldon Low and near the Weaver-hills. Brick earth of good quality is excavated in many places. The limestone marble peculiar to this county is valuable, but not of sufficient importance (to the engineer, at least) to require further notice.

Salop.

Shropshire contains, 1. *Coal* and *Trap*. 2. *New red sandstone*. 3. *Lias*. 4. *Silurian rocks*. 5. *Old red sandstone*. 6. *Magnesian limestone*. And, 7. *Carboniferous limestone*.

Shropshire presents abundance of materials to the notice of the engineer, but the limits of this Register will only permit the enumeration of the principal quarries.

In addition to the limestone near Oswestry and Chirk, a long range of stony strata (silurian) extends from Colebrook Dale, by Wenlock, to Ludlow. The most extensive lime works are at Lilleshall. Limestone quarries are worked at Steiraway; and lime and stone may be obtained in the Clee-hills, and, to a limited extent, in the south-western districts. At the eastern extremity of the Wrekin, and at some other lime works, a red lime of the hydraulic kind is made. The best lime of this county is made from the limestone of the lias beds. The principal sandstone quarries are at Grimshill and its neighbourhood, near Bridgenorth, Orton Bank, parish of Bettws, (stone slate,) on the south-western verge of the county; Coondon hill, (flagging,) Swinney quarries, near Oswestry; Bowden quarry, in the hundred of Munslow and Soudley, (flagging,) in the parish of Eaton and franchise of Wenlock. The principal hills are, 1. Lawley-hill, a kind of *granite* and *toadstone*. 2. Caer Caradoc-hill, *schistus*. And, 3. The Wrekin, chiefly composed of *reddish chert* and *quartz*. The Oswestry hills are a *coarse-grained sandstone*.*

* Lewis's Top. Dic. The student is also referred to an elaborate paper "On the Wrekin and on the great Coal Fields of Shropshire," by Arthur Aikin, Esq. Geol. Trans., vol. i., p. 191.

Hereford.

Herefordshire contains, 1. *Old red sandstone.* And, 2. *Silurian rocks.*

The greater portion of this county is occupied by the old red sandstone formation; this deposit, therefore, will now demand our attention. The *old red sandstone* consists of "beds of sandstones, conglomerates, clay, and concretionary limestone of various colours, but mostly red;" the subdivisions consist of quartzose, conglomerates and sandstones, coloured marls, concretionary limestones, and flag-stone series (Walker—map). This sandstone is generally of a dirty iron-red, or dark brown, occasionally passing into gray; it rests upon graywacke rock, into which it passes insensibly; it is micaceous, coarse-grained, and apparently constituted of abraded quartz, felspar, and mica; it alternates with argillaceous beds. Subordinate to it are some unimportant beds of limestone. In general appearance it approaches to the sandstones of the millstone grit series (Conybeare. "Geology of England and Wales," p. 362). The limestone beds below the surface near Snodhill castle, were originally worked for the kind of marble they afford; the marble (red and white) is said to have been in much estimation during part of the seventeenth century. In addition to the quarries of red sandstone* distributed through-

* It is observed, (Brande's Journal, vol. iii., p. 381,) that "the fitness of the different species of sandstone for the purpose of building, may, in a great measure, be judged of by immersing the specimens in water, each

out the county, an inferior kind, called "dunstone," is also quarried. Much of the limestone of this

being previously weighed, and all of one size, the excellence of the stone will be inversely to the quantity of water absorbed." A perfect analysis is, however, for many reasons, to be preferred. The most perfect one that has fallen under the notice of the author, is accurately described in No. 13 of the "Civil Engineer and Architect's Journal." It is a translation from the German of Dr. Buehner, professor in the University of Vienna. The following is the method, extracted from the "Journal:"—

"A piece of gray sandstone weighing 30⅝ ounces, was laid in distilled water for twenty-four hours; and on being taken out and weighed, it was found to have increased six grains, hardly two per cent., thus affording a good proof of its closeness of formation, and small power of absorption. The water in which the stone had been laid was evaporated to an ounce, and a yellowish residuum was obtained, which, on being subjected to reagents, was found to consist of sulphate of lime and sulphate of soda, mixed with organic matter. A piece of the sandstone was pulverized, and 100 grains of it treated with muriatic acid, and a partial dissolution effected by the development of carbonic acid gas. The remaining acid having been renewed by evaporation, the residuum of quartz sand was washed and cleaned with warm water, and found to weigh 57 grains. The muriatic residuum was subjected to nitrate of ammonia, whereby alumina was produced, with a portion of oxide of iron. It weighed, on careful trial, 3½ grains. The solution filtered from the aluminous precipitate was treated with oxalic ammonia, to produce deposition of the lime, which was exposed to the fire, to convert the oxalate of lime into carbonic acid gas, and by which 24 grains of carbonate of lime was produced. The fluid filtered from this was acted upon by phosphate of natron, and a precipitate of phosphate of ammonia and magnesia appeared, which, by heat, was reduced to neutral phosphate of magnesia, which was calculated as 13 per cent of carbonate of magnesia. The composition of the stone was, consequently,—

Quartz........	57
Alumina......	3.5
Carbonate of lime	24
Ditto magnesia..	13
Loss..........	2.5
	100

county being argillaceous, burns to lime of good quality.

Worcester.

The county of Worcester contains, 1. *Lias.* 2. *New red sandstone.* And 3. *Sienitic granite* (Malvern hills).

In Worcestershire, Bath stone and lias lime is extensively used; but the building materials in the immediate locality, consist of sandstone of the *new red formation,* and limestone of the lias and silurian beds. Lime and limestone is procured at South Littleton, Witly, and Huddington. Quarries of calcareous flagstone (lias) are worked in the vale of Evesham, in the parishes of Badsey, Three Littletons, and Prior's Cleeve. Lias limestone is also quarried in the vicinity of Stoke Prior. The Malvern sienite is occasionally used for metalling the roads. The bricks manufactured in the county are firm, and of a bright red colour.

Warwick.

The county of Warwick contains, 1. *Coal.* 2. *New red sandstone.* 3. *Lower oolite.* And, 4. *Lias.*

"From these results it was proved that the sandstone of Waakirchen was a good building material, and fully capable of resisting the effects of air and water, as its component parts were not liable to decomposition, and its texture did not admit the introduction of their mechanical force. It is evident that it is only by such trials that the true qualities of materials are to be ascertained, as mere mechanical action, or a trial of temperature, affords no criterion of the chemical constitution by which injuries of weather are caused."

Freestone quarries of the new red sandstone and lias formations are to be found near Warwick, Leamington, Kenilworth, Coventry, &c. Limestone quarries are also situated near Bearley, Grafton Court, Stretton, Princethorpe, Upton, Harbury, Wilncote, Bidford, Newbold-on-Avon, and on the borders of Oxfordshire and Leicestershire. Blue flagstone (lias) is quarried at Bidford and Wilncote. A ferruginous sandstone is to be met with at Oldbury and Merevale.

Other kinds of stone are used in this county, namely, Gornal stone, (prices, 1838,) 1s. 4½d., per cube foot; Wharton stone, 1s. 3d.; blue stone, for paving, 10d. to 1s.; Bath stone, 2s. 9d.; Painswick stone, 3s. Bricks, (red,) 23s. to 26s. per thousand, according to distance.

Northampton.

The county of Northampton contains, 1. *Lias.* 2. *Lower oolite.* Building stone is raised at Brackley Kingsthorpe, near Northampton. Stone slate, or thin flagging, is dug at Collyweston, near Stamford. The oolite called "Ketton stone" is found near Stamford, and quarried there. The lime of this county is mostly of the lias kind, and is quarried on the western side.

The older buildings of the county have been constructed with the rusty sandstone from the upper beds of the lower oolitic series. The bricks are of two kinds, red and light-coloured; many of the latter are procured from the borders of Huntingdonshire.

Huntingdon.

Huntingdonshire contains, 1. *Green sand.* 2. *Upper, middle,* and *lower oolite.*

A considerable quantity of the stone used in this county for the finer purposes of building, is raised from the Ketton quarries, (oolitic,) near Stamford;* the inferior kinds of stone (car stone, or quern stone) are procured from the iron sand deposit, (one of the beds between the chalk and oolites,) and the sandy beds of the lower oolite. The brick earth of this and some other of the midland counties is micaceous, and belongs to the beds between the chalk and oolites. The bricks are of two kinds, light and red coloured.

Cambridge.

Cambridgeshire contains, 1. *Chalk.* 2. *Green sand.*

The principal building materials supplied by this county are, 1. An excellent light-coloured brick, made from the gault or clay. And, 2. A lime made from clunch, a calcareous substance found in large masses, principally in the parishes of Burwell and Isleham. It is also used as a fire-stone.

* A good example of its application in the Metropolis, may be seen at the new St. Dunstan's church, Fleet-street.

The geology of Cambridgeshire is, in many respects, very interesting. An able paper on the subject, by Professor Hailstone, is to be found in the third volume of the Geological Transactions (1816,) from which we give the following interesting extract, relating to the beds provincially termed "clunch:"—

"These beds contain no flints, but, not uncommonly, dispersed masses of the radiated pyrites, globular or kidney form. It is considerably harder than the common chalk, and its colour is usually some shade of gray. It is well known in this county under the name of '*clunch*,' and is the material from which the best lime is burnt. Some of the beds are hard enough to serve the purpose of building stone, and are quarried and shaped into blocks for that purpose. It also endures the fire well, and, like the Reygate stone in London, is much esteemed for the backs of grates, and other similar applications. This stone is dug in the greatest quantities at Reach, a small hamlet in the parish of Burwell, situated on the skirts of the fen country, precisely where the Devil's Ditch terminates in that direction. The excavations at this place are immense. Clunch, when burnt, affords a lime in such universal esteem, that the crude material is sent from hence, for that purpose, as far as Peterborough, and other distant places within reach of the water carriage of that level district."

The stone employed in some of the new buildings at Cambridge is from the Whitby, Portland, Painswick, and Ketton quarries.

Norfolk.

The county of Norfolk contains, 1. *Crag* and *diluvian.* 2. *Chalk.* 3. *Green sand* and *upper oolite.*

In this county the materials, with the exception of brick, must be obtained from the adjoining counties, or brought coastwise from the principal stone quarries of England.

The most ancient buildings in the counties of Suffolk and Norfolk, are constructed with flint, or boulder walls, and limestone, (oolitic,) or brick quoins.

Suffolk.

The county of Suffolk contains, 1. *Crag.* 2, *London* and *plastic clay.* 3. *Chalk.* And, 4. *Green sand.*

With respect to building stone and brick, the observations above will apply also to this county; but the local deposit called "*crag,*" requires a passing notice.

The extent of this marine deposit is not yet ascertained. (Conybeare). The districts with which we are best acquainted are the eastern parts of Norfolk and Suffolk, extending to Walton Naze, in Essex. This deposit consists of an heterogenous mass of matter, namely, sand, gravel, blue and brown marl, and broken shells. In some places it assumes the character of a soft, stratified rock, composed almost entirely of corals, sponges, and echini; in other parts it consists in alternations of sand and shingle, destitute of organic remains, and more than two hundred feet in thickness, as in the Suffolk cliffs, between Dunwich and Yarmouth. (Lyell). The more consolidated portions are occasionally employed for ordinary building purposes, and quarries of it are worked at Southwold, and on the southern bank of the river Orwell, in Suffolk (Conybeare).

The bricks manufactured in Suffolk are from a light clayey loam; they burn to a very light colour, and are much used for facings.

Essex.

The county of Essex contains, 1. *London clay.* And, 2. *Plastic clay.*

The materials supplied by this county chiefly consist of bricks of good quality, and septaria, or English cement stone, procured from the Harwich cliffs. The peculiar nature of the London and plastic clays, and the composition of septaria will be found duly noticed in the 1st and 5th chapters of the "Treatise."

Much of the lime used in this county is procured from Purfleet and Grays; it is, also, occasionally brought from Dorking, Greenhithe, &c., on the other side of the river Thames. The London clay constitutes a very large part of the soil of Suffolk, nearly the whole of Essex, including Hainhault and Epping forests, quite to the sea, the whole of Middlesex, and portions of Berkshire, Surrey and Kent (Phillips). The London clay is often found immediately under the alluvial soil, but is more frequently covered with a diluvial coating of gravel, of considerable thickness.

Hertford.

Hertfordshire contains, 1. *Plastic clay.* And, 2. *Chalk.*

The soils of Hertfordshire are chiefly gravel, clay, and chalk. The gravel tract lies in the south eastern part of the county, and contains 17,280 acres. The chalk district, extending along the whole border of the northern part of the county, contains 46,720 acres. The clay districts are situated, one on the southern, and the other (the largest) on the eastern side of the county. The materials of this county are bricks of good quality and chalk lime.

Bedford.

Bedfordshire contains, the *Upper, middle,* and *lower oolitic series,* bounded by *green sand.*

With the exception of some very inferior sandstone belonging to the upper beds of the oolitic series, the strata of Bedfordshire does not supply stone fit for the purposes of masonry. Its contiguity to the counties of Buckinghamshire, Oxfordshire, and Northamptonshire will, however, enable the architect or engineer to draw his supplies with facility, should the materials peculiar to these counties be considered suitable for the intended works. Lime and bricks may be procured on the borders of Hertfordshire.

Buckingham.

Buckinghamshire contains, 1. *Plastic clay.* 2. *Chalk.* 3. *Green sand.* 4. *Upper, middle,* and *lower ooolite.*

The first important development of the oolitic series of rocks is to be found in this county, and from which the whole of the series may be successively traced, namely, through Oxfordshire, Berkshire, Wiltshire, Dorset, Somersetshire, Gloucestershire, &c.

The Aylesbury or Portland limestone first makes its appearance in that part of Buckinghamshire which lies west of the Grand Junction Canal. It forms the central and lower regions of a chain of hills capped by the *Iron sand:* at Brill Hill, the Portland beds rise nearly to the summit, having at that spot only a slight covering of iron sand (Conybeare). The Aylesbury limestone quarries contain a series of beds of unequal quality,—namely, a fine-grained white oolite, a loose granular limestone, of earthy aspect, and of various shades of yellowish gray; and, more rarely, a compact cretaceous limestone, having a conchoidal fracture, (Conybeare). The more oolitic variety of the upper series (Portland) and the argillaceous variety above the Portland (Purbeck) is described in the account of the Portland and Purbeck quarries.

In addition to the limestone quarries already mentioned, marble was formerly quarried in this county, but it was found incapable of retaining a good polish, or withstanding the effects of the atmosphere; its use has, therefore, been discontinued. (Lewis Top. Dic.)

Oxford.

Oxfordshire contains, 1. *Green sand.* 2. *Chalk.* 3. *Upper*, *middle*, and *lower oolite*, and is adjacent to the *lias* formation.

The oolitic system is more extensively developed in this than in the former county, but the materials drawn from the *middle* or *coral rag* division of the oolitic series (the one under notice), are of a very inferior description, being highly objectionable for the finer purposes of masonry, and the different beds (near Oxford and in Wiltshire) are exhibited in the following progression, commencing from the top. 1. A *calcareous freestone* of tolerably close texture, more or less oolitic, but full of comminuted shells, and passing into beds of a large oviform grain, forming the "*pisolite*" of Mr. Smith. The colour of these beds is of a yellowish white, becoming palest in the most oolitic, and passing occasionally into shades of light gray. This bed affords a tolerable material for building, but is objectionable on account of its being traversed by lines of division oblique to the plane of stratification (Conybeare).

Much of this stone has been used in the collegiate and ecclesiastical buildings of Oxford, where the peculiar defect of its stratification may be observed in the scaling or peeling off of the outer surface, after exposure to the weather: this stone contains a considerable quantity of sand. The *coral rag* immediately follows the bed of stone already described, and consists of a loose rubbly limestone, made up of several species of aggregated and branching madrepores; it is used for lime and mending the roads (Conybeare).

The beds of coral rag form an elevated platform, rising on the south-west of Otmoor, and supports the still higher ridge (Portland beds and iron sand) which constitutes the summit of Shotover Hill. The quarriesare situated on this platform, and a e very numerous; the principal quarries are those of Headington, two miles east of Oxford (Conybeare).

The several clays* which divide the oolitic series, require no further observation, than that they afford to the engineer or builder an opportunity of procuring a supply of bricks throughout a very extensive district in which limestone is abundant.

Gloucester.

Gloucestershire contains, 1. *Lower oolite.* 2. *Lias.* 3. *New red sandstone.* 4. *Mountain limestone.* 5. *Old red sandstone.* And, 6. *Coal.*

The upper beds of the lower oolitic series furnish a considerable quantity of lime and limestone, which is much used in the Cotswold district. The principal upper beds of this series (in Gloucestershire) are the *corn-brash* or *corn-grit, forest marble, Stonesfield slate,* &c. The *Corn-brash* is a loose rubbly limestone, of a gray or bluish colour, especially near the superincumbent clay, but on the exterior brown and earthy; it rises in flattish masses, rarely more than six inches thick. It is not fit for any purpose, excepting for lime and the repair of the highways; but at Malmesbury, where it is thick and solid, it is much quarried for building. It may readily be discovered by the red soil which constantly attends it† (Conybeare).

* The Oxford or *clunch clay* is a dark blue adhesive clay, containing *septaria*: it underlies the limestone already described, separating the inferior or lower oolite from the former.

† We may here observe, that beds of forest marble are worked near Woodford (Bucks), and at Marsh Gibbon, Ambrosden, Merton, and

The greatest thickness of the lower oolitic beds,* is exhibited in the continuous chain of hills, known as the Cotswold hills. Many quarries are worked on these hills, and others are situated near Badmington Park. In Gloucestershire, the *great oolite* always crowns the brow of the escarpment of the hills. The loftiest point of these hills, is Cleeve hill, near Cheltenham. (Conybeare).

The limestone of the Forest of Dean, Langhope, and adjacent places is quarried for building, but the lime is inferior to that made from the vast limestone beds of the southern extremity of the county. (See Chapter iv. of the "Treatise").

Freestone is raised from the Cotswold quarries, and paving stone is raised at Frampton, Cotterell, Winterbourne, Iron Acton, Mangotsfield, Stapleton, and in the Forest of Dean; the latter place supplies millstone, and other grits of excellent quality. Stone slate may be procured on the Cotswolds.

Wiltshire.

Wiltshire contains, 1. A small portion of *plastic clay*. 2. *Chalk*. 3. *Green sand*. And, 4. *The oolitic series*.

Bletchingdon (Oxon). From the latter quarries, the stone for the pillars in the inner quadrangle of St. John's college, Oxford, were procured (Conybeare). The calcareous slate quarries of Stonesfield, near Woodstock, are situated in the valley immediately on the south of Stonesfield village. The forest marble is composed of little else but a mass of shells (Ibid.)

* The limestone of the Cotswold district, is arranged in beds, from two

The limestone beds of the upper division of the oolitic series, (Purbeck, Portland, and Kimmeridge beds,) are extensively quarried in Wiltshire. The principal quarries are at Fonthill, Tisbury, Chicksgrove, and Chilmark.

Berkshire.

Berkshire contains, 1. *Plastic clay.* 2. *Chalk.* 3. *Green sand.* 4. *Upper* and *middle oolite.*

The building materials of this county are chiefly bricks and chalk lime. The brick earth belongs to the *plastic clay* formation, and is excavated in considerable quantities at the Catsgrove fields, near Reading, and other large quarries of brick earth on Saint David's hill, west of Reading, where this formation is about forty feet in thickness. (Buckland's Geology Trans. vol. 4.) The red clay of Reading, on the north of the Hog's-back, and at East Horsley, is perfectly identical with that of Meudon in France; nor has this colour been found equally intense in any other clay. The bricks made of this clay are of a bright Roman ochre colour. (Conybeare, Geology of England and Wales, Notes.)

The quarries of chalk in the neighbourhood of Reading, is extracted largely from under the sands

to ten inches in thickness. In this district, the fields are not divided by hedge rows, but by walls built with the *stone brash,* without mortar.

and clays, by means of shafts and levels, to be burnt into lime. (Buckland.)

Siliceous grit stones, (called *sarsden stones* and the *gray weathers,*) are distributed in a loose state over the Berkshire and Wiltshire downs, and apparently, do not belong to the contiguous strata.

Middlesex.

The county of Middlesex contains, *London* and *plastic clay.*

A particular account of the London clay will be found in the first Chapter of this Treatise, accompanied by some sections of the strata. The first fifteen or twenty feet of the clay stratum is of a chesnut colour, but at a greater depth, changes light blue; and finally, passes into a dull black. The upper stratum is generally used for *bricks,* and the lower ones for *tiles ;* either will answer the purpose for brick making, but the blue clay will burn to brick of a deep red colour. This may, however, be prevented by the admixture of a certain proportion of chalk with the clay, (in a pug mill,) which will then burn to a bright straw or ochre colour. The number of brick fields in Middlesex is considerable. Immense quantities have been made at Hackney, Kingsland, City-road, Hounslow, Brentford, and Cowley, near Uxbridge. The bricks from the latter field are much valued, on account of their bright yellow colour, and general good quality.

Surrey.

The geological formations which are developed in this county, consist of the *chalk, green sand,* and *Wealden.*

The chalk division comprises, 1. The *upper chalk.* 2. The *lower chalk.* And, 3. The *chalk marle.* The green sand division contains, 1. The *upper green sand,* containing sand intermixed with green particles, (chlorite), indurated marle, and calcareous sandstone. 2. The *gault,* consisting of a series of clay beds passing into calcareous marle. 3. The *lower green sand,* containing beds or strata of gray, yellow, and greenish sands, ferruginous sands, sandstones, clays, &c., and siliceous limestones. The Wealden contains, 1. *Weald clay.* 2. A series of *sand, sandstones, clays,* and *calcareous grits* passing into limestone.

The material furnished by this county, for architectural or engineering purposes, is now chiefly confined to the supply of lime for the use of the metropolis, and adjacent places; the best kind is that manufactured from the indurated *chalk marle.*

The kilns are situated at Dorking, Croydon, Hallam, Riddlesdown, Merstham, Guildford, Reigate, Sutton, Carshalton, &c.; and the principal quarries are those of Croydon, Dorking, Sutton, Epsom, Leatherhead, Bookham, Effingham, Horsley, Claudon, Stoke, Guildford, and Puttenham, on the northern side of the Downs; and those at Godstone, Catterham, Reigate, Merstham, Buckland, and Betchworth, on the southern side.

The upper green sand division supplies a sandstone which was formerly employed in important buildings, and we are informed that a patent of Edward III. exists, authorizing the working of the quarries at Gatton and contiguous places, for the supply of stone for Windsor Castle. Henry the Seventh's chapel at Westminster, and the church at Reigate, &c., also furnish examples of its employment as a building stone.

The older quarries are now abandoned, and the stone has been for some years past, procured at Merstham. According to an examination made by Mr. Webster,* its constituents appear to be siliceous sand and mica, cemented by an earthy carbonate of lime, with a very small quantity of dark green particles. Mr. Webster observes, that "this latter substance is not so abundant as is generally the case in the green sand formation, a glass being necessary to observe it, but it is never totally wanting."

The stone from these quarries is very unfit for the purposes of masonry, in consequence of the carbonate of lime contained in it being impregnated with earthy matter. If used without the precaution of drying, it soon falls to sand, but it may be employed with advantage as a *fire stone,* or in such situations as may secure it from the attacks of the weather.

* See Geological Transactions, vol. v, part 2, p. 355.

The calcareo-siliceous grits of the green sand formation vary considerably in their state of induration: stone has recently been procured from Leith Hill, about six miles from Dorking, of a much better quality than the Merstham stone. The villa of Mr. Spottiswoode, and some other buildings lately erected near this place, may be referred to as practical examples of its application. The bricks manufactured in the county of Surrey are mostly of a red colour, but are much inferior in quality to the *gray stock* of Middlesex, Kent, and other districts covered by the London clay.

Fire bricks and tiles of good quality are manufactured at Ewell, the materials being derived from the beds belonging to the plastic clay formation, and, occasionally, blocks of an indurated sandstone are obtained from the deposit, called Bagshot sand (upper marine formation), sufficiently firm for the purposes of masonry. These two latter deposits are however, but feebly represented in the county of Surrey.

Kent.

The maritime county of Kent contains portions of the following formations:—1. *London clay.* 2. *Plastic clay.* 3. *Chalk.* 4, *Green sand.* And, 5. *Upper oolite.*

Building materials of the following kinds may be obtained with facility from various parts of the county:—1. From the London clay, Septaria, or English cement stone, from the Isle of Sheppy; 2. From the chalk, chalk marle; Lime-kilns and chalk-quarries are situated at Greenhithe, Northfleet, Gravesend, Rochester, Chatham, Chevening, Dover, and at other places on the coast; 3. From the green sand, Kentish rag: the western part of the county of Kent is the principal depository for the Kentish rag-stone. The rock forms the sub-stratum of a soil, called *hassock*, or *stone shatter*, which is a mixture of sandy loam and fragments of the rock; blocks of stone, of considerable dimensions, are frequently obtained within a few inches of the surface. This stone belongs to the *lower* green sand formation; it is a calcareo-siliceous grit, possessing a considerable degree of induration, and is much used in various parts of the county; it is a substantial building stone. The beds between the chalk and oolites also contain sandstones of various kinds: at Penshurst, a tender ferruginous sandstone (from the iron sand formation) is quarried for building; and in the neighbourhood of Tunbridge, a hard quartzose ferruginous stone is extensively employed; the latter is esteemed as a good weather stone; 4. From the upper oolite. The argillo-calcareous beds of the oolitic formation, in the county of Kent, are occasionally worked for the limestones they afford, but the stone is not sufficiently durable for masonry. The limestones of this series are subsequently noticed by the names of Petworth and Purbeck marble. The plastic clay formation supplies brick earth and sands of excellent quality, immense quantities of stock and other bricks have been supplied to the London builders from the fields at Erith, Lewisham, Bromley, &c.

Sussex.

The maritime county of Sussex contains, 1. *Chalk.* 2. *Green sand.* And, 3. The upper division of the *oolitic series.* The materials supplied by the chalk and green sand formations, are similar to those found in the counties of Surrey and Kent. Lime from the chalk marle may be procured at Lewes, Brighton, Shoreham, Worthing, East Grinstead, Rye, &c., and at many other places in the chalk district. The *weald* of Sussex contains *limestone, limestone marble, sandstone,* and *ironstone.* The ironsand deposit (one of the beds between the chalk and oolites) is laid open at Horsham, Hastings, &c., and produces a very durable sandstone.

The groin work at Battle Abbey is of *freestone* belonging to this series; it is also used for copings, and in the neighbourhood of Horsham this formation yields flagstones for paving.* The tender limestone called Petworth marble, is procured from the weald clay in the neighbourhood of Petworth; its texture is earthy and friable; the stone is therefore unfit for exposure. The quarries, (with the exception of those

* Conybeare, Geol. of England and Wales. Notes, p. 137.

near Kirdford and North Chapel, in Surrey) are now abandoned.

*Hampshire.**

The county of Hampshire contains, 1. *London clay.* 2. *Plastic clay.* 3. *Chalk.* And, 4. *Green sand.* The chalk hills cross the county from east to west. The chalk is quarried extensively throughout this district, and near the coast; two kinds of lime are produced, one from the white and the other from the gray chalk; the lime from the latter is equal to the Dorking lime, and is much used. The plastic clay and green sand formations of this county and the Isle of Wight, produce sandstones and conglomerates available for building.

The following varieties are ordinarily used as building materials. 1. Between Milton and Christchurch a *ferruginous sandstone.* 2. A *porous grit* or *quarry stone,* obtained from the northern foot of the Downs, and found in detached masses. 3. A liver coloured *ferruginous sandstone* from the southern part of the island. 4. A *calcareous grit* from the marl pits eastward of Staples heath. 5. A close *gray limestone* procured from the northward of Arereton downs. 6. A marly variety of *freestone.* And, 7. *Conglomerates,* used for paving and flooring; these are procured in large quantities from near Sandown

* The Isle of Wight contains the following formation. 1. *Fresh water beds.* 2. *Plastic clay.* 3. *Wealden.* 4. *Green sand.* And, 5. *The Chalk.*

fort. Brick earth and potter's clay of excellent quantity abound in various parts of the county, and the bricks from Southampton are a substantial and elegant material for building.

Dorsetshire.

The county of Dorset contains, 1. *Plastic clay.* 2. *Chalk.* 3. *Green sand.* 4. the *Oolitic series.* 5. *Lias.*

A considerable portion of the county of Dorset is occupied by the chalk formation and its accompanying beds, above and below. The information already given, concerning the beds of this formation, and the materials which they supply, will be found sufficiently indicative of the resources of such part of this county as is occupied by these formations.

To the engineer or builder, the most important part of the county of Dorset is that occupied by the Isles of Portland and Purbeck, the grand depositories of the finer architectural materials of this county. The island of Portland consists of *black shiver,* covered with limestone (Middleton). The north end of it is about 400 feet above the sea, and that of the south is supposed not to exceed 100 feet in height.

The quarries are on the eastern and western sides of the island, and the cap of stone averages 100 feet in thickness. The stone is in layers, varying in thickness, with partings of clay, shiver, flint, &c. The best quarries are at Kingston, Worth, Langston, and Swanwich; other quarries are situated at Wareham, Morden, and near Dunshay. The quarries are also known by various names, such as Waycroft quarry, Goslings quarry, Winspit quarry, Maggott quarry, &c., and Lias is extensively quarried at Lyme Regis. The quality of the Portland and Purbeck stone, Lias, &c., will be found described in the 5th chapter.

Somerset.

The county of Somerset contains, 1. *Green sand.* 2. *Lower oolite.* 3. *Lias.* 4. *New red sandstone.* 5. *Coal.* 6. *Mountain limestone.* 7. *Clay-slate, graywacke,* and some small patches of chalk.

The lower division of the oolitic series is well developed in this county, and extensively worked for the supply of the well known Bath stone. The quarries around Bath are very numerous; the most important are Box, Coombe down, and Farley down, near Bath; and at Doulting, near Shepton Mallet, are some celebrated quarries of valuable freestone. The *lias,* which forms the base of the oolitic series, is extensively quarried in this county; a considerable quantity is procured from the quarries near Bath and at Watchet, St. Decuman's, Glastonbury, Somerton, Shepton Mallet, &c.

Mr. Horner, in describing the Somersetshire lias* observes, that "all the strata of this limestone,

* Geology of the south-western part of Somersetshire, by Mr. Horner, Geol. Trans., vol. 3.

though externally very similar, are not of the same mineralogical composition, for they have very distinct properties. Some of them yield a lime which possesses, in a most eminent degree, the property of setting under water: these are generally the thinnest strata, are of a light blue colour, and compact earthy texture; on each surface of the stratum, and at the joints, the colour is changed to a light brownish yellow or cream colour, which is of different thicknesses; in some places extending so far into the interior of the stratum, that the blue colour is nearly obliterated. The other variety of the limestone is of a much darker colour, but is most particularly distinguished by the strong fœtid smell it gives out when struck by the hammer, and when it is burning in the kilns. It is always in thicker strata than the other variety, and abounds in organic remains; it is also very much penetrated by pyrites in many places. These fœtid strata have much less the property of setting under water, and are best adapted for agricultural purposes, for which the other are very unfit."

The first mentioned variety of limestone is by the quarry-men called blue lias or building lime, and the latter, black lias or ground lime; the strata is generally less adapted for building lime the lower they go. The limestone on the eastern side of the Quantock hills, is chiefly of the crystalline kind, but of variable quality, and accompanied by the red sandstone and graywacke of the county. The principal Quantock quarries in the south-western part of

Somersetshire, may be enumerated as follows:—Binfords, Westcot, Treborough,* Leigh, Doddington, Triern farm, Ely green, Cannington park, &c. (Horner.)

A very fine grained granite, provincially called "*pottle stone*," was formerly quarried at the foot of a hill, a few miles north-east of Taunton.† The quality of the various kinds of Bath stone is fully described in the 5th chapter of the Treatise.

Devonshire.

The county of Devon contains, 1. *Green sand.* 2. *New red sandstone.* 3. *Millstone grit.* 4. *Granite, trap,* &c. 5. *Clay-slate.* And at Chudleigh, a portion of *clay.*

The granite quarries are situated on the elevated tract called Dartmoor, in the south-western part of the county, and is surrounded by a district of argillaceous slate. The transition slate occupies the northern part of the county, including Exmoor. The red sandstone occupies the lesser tracts of the county, and skirts the base of the last mentioned district, extending north-eastward into Somersetshire, and westward as far as Hatherleigh. The green

* At Treborough, a very excellent roofing slate is to be procured. Horner, Geol. Trans.

† There are extensive limestone quarries in the neighbourhood of Bristol.

sand constitutes the largest portion of the hills in the south-eastern part of the county (Lewis).

The extensive quarry of Heavitree, (red sandstone formation,) is situated about a mile and a half from Exeter, on the road to Honiton. The stone is a stratified conglomerate, compact and tenacious, but passes into a tender and friable sandstone. Dr. Berger * (to whom we are indebted for this information) states, that owing to the presence of a considerable quantity of calcareous particles in the stone, it might be very easily mistaken for limestone; it also contains particles of other rocks.

Poucham quarry is about two miles N. N. W. of Exeter. The rock is an amygdaloid, the nodules of which are chiefly calcareous, and much stained with the oxide of iron.

There are three or four quarries at Thorverton, seven or eight miles north of Exeter, in the same amygdaloidal rocks; but the composition of the rock varies considerably in different places.

The limestone of the cliffs from Stonehouse Pool to Catwater and Mount Batten, is formed of a compact limestone, which breaks into large flakes with a semi-conchoidal fracture. It is of a yellowish white colour, and, when quarried, is blasted with gunpowder. On the side of Catwater this limestone is of a bluish colour, and crystalline grain, but intersected by veins of calcareous spar, in many places.

* Dr. Berger "On the Physical Structure of Devonshire and Cornwall." Geol. Trans. vol. i., p. 101.

The transition limestone of Plymouth has already been sufficiently described; there is, however, a quarry of it on the left bank of the Plym, belonging to Lord Borringdon. The stone is of a blackish brown colour, containing rhomboidal plates of calcareous spar. It is remarkable, that in the same stratum, it suddenly assumes all the characters of a shining slate, and, in this state, effervesces less briskly with acids.

Cornwall.

The county of Cornwall contains, 1. *Millstone grit.* 2. *Granite.* 3. *Trap rocks.* 4. *Clay* and *graywacke slate.*

The greater part of the county of Cornwall is occupied by the Killas, a coarse argillaceous schist, which is found in every part of Cornwall with the exception of the spaces occupied by the granite.

The granitic districts are four in number. The first is nearly bounded by the church towns of Northill, St. Neot's, Blislind, St. Breward, and St. Clether; the second, by Llanlivery, Roach, St. Dennis, St. Stephen's, St. Austell, and St. Blazey; the third, by those of Constantine, Crowan, Redruth, and Stithians; the fourth occupies the western extremity of the county, from St. Paul and Tennor, to the Land's End. There are also four small spots of granite; the first between Calstock and Callington, the second east of Redruth, a third west of Breage and St. Michael's Mount.

There are some thin beds of crystalline limestone near Padstow, in the parishes of Carantoe and Lower St. Columb, also between Liskeard and the Tamar. Beds of china and other clay are found in various parts of the county (Lewis).

It would appear that the granitic rock of Calstock possesses a certain degree of stratification, for in working the stone, the quarrymen split it with a great number of wedges. The *pool* holes are first sunk with the point of a pick, about four inches apart, according to the size and supposed strength of the stone.

Mr. Smeaton states, in his "Narrative," that he has frequently seen, in this county, "posts of granite twelve feet long, and not above eight inches square, used instead of wood, for the mere purpose of supporting a *Hovel*." "But it is to be constantly observed," he continues, "that the strength of the stone on each side of the bisection is so managed, as to be nearly equal, otherwise the *split* will constantly encroach on the weaker side."

Isle of Man.

The Isle of Man contains, 1. *Mountain limestone.* 2. *Clay slate.* And, 3. *New red sandstone.*

In this island there are several quarries of valuable freestone. The dark gray or black marble so much employed in the time of Wren, was procured from this island. The external steps of St. Paul's cathedral were, it is stated, procured from these quarries.

The intermediate strata of the formations which we have enumerated, are of almost infinite variety, but when considered as substrata for building upon, may be arranged in the following order—Bog earth, Clay, Gravel and Sand. We shall have occasion in the course of the following Treatise, to point out in detail, the precautions necessary to be adopted, when building upon soils so varied in character; in this place, however, few general remarks appear necessary. The most recent and successful instance of forming a foundation on moss or bog, is exhibited on the Liverpool and Manchester railway. Chat Moss is a vast morass, situated on the north side of the turnpike road, leading from Manchester to Liverpool, containing about 6,000 statute acres, the entire extent covering about twelve miles. The composition of this bog is identical with the red bog of Ireland, and the white flow moss of Scotland. The vegetable mass retains the water similar to a sponge. Every fibre contains a portion of water, and the tenacity with which it continues to hold it, gives to the whole a semi-fluid quality, not always free from danger to the neighbouring land. The depth of this bog is, in some parts, forty feet, resting of course, on what was the original surface; the general drainage was therefore, accomplished with little difficulty. The Liverpool and Manchester railway crosses Chat moss, in a line of nearly five miles in length. In forming the embankment at the eastern boundary, (twenty feet in height,) an immense mass of earth was thrown in and disappeared, before any thing like a palpable foundation for the road could be obtained.

In forming the embankments on the moss, 677,000 cubic yards of bog earth were dug up, and the water being ejected, so as to render it solid and fit for the purpose, its quantity was reduced to about 277,000 cubic yards. The embankment across *Parr's Moss*, three quarters of a mile in length, was composed of 144,000 * cubic feet of clay and stone, procured from the Sutton inclined plane.†

In building upon a substratum of London clay, it is highly necessary that the excavations should be continued *below*, what may with propriety be called the *weather line*. The London and other similar clays, are remarkable for the natural fissures which they contain, many of which are either wholly or partially filled with calcareous spar or sulphate of barytes, but in exposed situations, and in the summer season, many additional cracks become visible, and extend to a considerable depth. On an atmospherical change taking place, the fissures close, and the clay swells; the enormous pressure thus exerted by the

* Qy. yards?

† See "Wheeler's History of Manchester," 1836.

water contained on or below the substratum, will effectually disturb the equilibrium of buildings of moderate weight. Great attention is also required in passing from one stratum to another. In all cases of soft soils, the upper stratum should be either wholly cut through, or a sufficient depth of the natural soil should be left under the foundations, equal to bear the superincumbent pressure. Great danger is, therefore, to be apprehended in breaking the upper crust of a new stratum; if a new stratum is to be passed into, the excavation should be carried down to the more consolidated part. Should the substratum consist of a gravelly deposit, its nature and extent should if possible be ascertained previous to the commencement of the works, for strata of this description offers more remarkable phenomena to the notice of the engineer than any other.

It should be remarked, that although sands and gravel "faithfully represent those rocks from which they are derived," yet vast accumulations of gravel are deposited in many parts of the country, derived from rocks whose localities are far removed from those deposits. The post-diluvian aggregations, notwithstanding their great thickness in some places, are generally loose and shingly in their nature, and if adopted as a substratum for important works, require considerable artificial consolidation. In the vicinity of London, the gravel chiefly consists of rolled flint pebbles, probably derived from the chalk formation; but to the westward of the metropolis, the gravel under the alluvium is intermixed with a considerable quantity of *quartzose pebbles.*

It appears that the nearest quartzose rocks in *situ*, are those which constitute the Lickey range, near Bromsgrove, in Worcestershire. The vast ruin of those rocks has been traced* from their source, through Warwickshire, Oxfordshire, and the valley of the Thames downwards, towards London.

The gravel pits of Hyde Park and Kensington, contain a considerable quantity of quartzose debris, much of which is either in a state of comminution or decomposition. Deposits of this nature are easily identified by their loose structure and heterogeneous composition, while the real diluvian is mostly held together with a calcareous or ferrugineous cement, in considerable stratified beds.

The gravel of Lichfield is an example of a decomposing strata, well worth the attention of the student or engineer.†

Foundations upon sandy beds have been formed with considerable success. The erection of the light house upon Spurn Point, by the late Mr. Smeaton, is a useful example to the student, and shews that a firm sand will resist the insertion of piles after the depth of nine feet.‡

* See Dr. Buckland on this subject, vol. 5, part ii. Geolg. Trans.

† See Aiken on the Lichfield Gravel, Geol. Trans. vol. 4.

‡ See Smeaton's "Narrative."

The bridge over the estuary of the Lary, at Plymouth, is another successful example of building upon a sandy bottom. An excellent memoir on this subject will be found in the first volume of the "Trans. Inst. C. E."

Piling on *alluvial sands*, *low situations*, or on the *banks of rivers*, is highly objectionable, inasmuch as the absorption of water from above, or the upward filtration from an adjacent retentive stratum, combined with the influence of the tides, will most certainly subject the buried timber to those changes of temperature, and alternate dryness and moisture, which are so destructive to even the most compact and solid materials.

The following materials have been carefully examined and experimented upon. The results will be given in the course of the Treatise:—

Strata above the Chalk.

From the London clay and sands above:—Highgate and Bagshot sands, and septaria from Primrose-hill.

From the chalk: — Merstham and Dorking chalk marle; lime made from ditto.

From the Green sand:—Sandstone from the Merstham and Leith-hill quarries, Surrey; Kentish rag from the Weald.

From the Weald clay:—Specimen from Petworth, and specimen of Purbeck from the foundation of Old London Bridge.

From the Iron sand:—Sandstone from Penshurst, Kent, and Sandy, Bedfordshire.

Oolitic series, upper division:—Limestone from Aylesbury, Purbeck and Portland.

Middle division:—Red sandstone from Weymouth; Limestone from Chippenham, (Wilts,) and Lincoln.

Lower division:—Limestones from Ketton, (Northamptonshire), Coombe Down, Lansdown, Farley Down, and Box quarries, Bath, (Somerset), Stamford, (Huntingdonshire,) and Newport Pagnel, (Bucks); Sandstones from Northampton, Fenny Compton, Blisworth and Daventry.

Lias:—From Bath, (Somerset,) and from near Stoke Prior, Worcestershire.

New red sandstone:—From the Liverpool tunnel, and Oldbury, (Warwickshire), Tamworth, (Staffordshire), and Runcorn, (Cheshire).

Magnesian limestones: — From Brodsworth, Ripon, Doncaster, Quarry Moor. (Yorkshire), Durham and Nottingham.

Transition limestone:—Plymouth.

From the carboniferous series: — Limestones

from Chesterfield and Matlock, (Derbyshire,) Roach Abbey, Richmond, (Yorkshire,) Dost-hill quarries, (Staffordshire); Sandstones—Warsall stone, Rainton quarry, Pateley stone, Moor stone, Bramley Fall, Mexborough stone, Holton stone, Kirby stone, Leeds flagging, Idle flagging, and Sheffield stone, (Yorkshire), Heddon quarry, (Northumberland), and Stokes quarry, Amington, (Staffordshire.)

Granites, Sienite: — From Great Malvern. Dublin, Hayter, (Devonshire), Constantine, (Cornwall), Aberdeen, (Scotland), and Sandstone from the Cragleith quarries, (Scotland).

NOTE.

The Author takes this opportunity of correcting an error (see p. 64) respecting the state in which the iron shoes of the piling, belonging to the old London Bridge, were found.

Upon communicating with the parties who purchased the old metal, (Messrs. Weiss, of the Strand) those gentlemen furnished a polite communication, of which the following is an extract:—"Your informant has fallen into an error with regard to the iron from the piles of old London Bridge. It was not converted into steel, although it had very peculiar qualities, namely, toughness in a very great degree, and the shoes, when struck, gave out a ringing sound like a bell, the result, we suppose, of its tenacity. It makes a very hard steel, which is not only beneficial to us as instrument makers, but is in request as gravers for engraving on steel, copper, and mezzotint."

CHAPTER I.

PRELIMINARY OBSERVATIONS.—REMARKS ON THE GENERAL ORDER AND SUCCESSION OF STRATA.—ON THE EXAMINATION OF STRATA, AND DESCRIPTION OF BORING APPARATUS EMPLOYED FOR THAT PURPOSE.—NATURE OF STRATA ABOVE THE LONDON CLAY.—SECTIONS OF STRATA IN THE NEIGHBOURHOOD OF LONDON, &c.

THE great importance attached to the study of the Science of Architectural Construction, and the vigour with which analytical investigations on this subject have of late years been followed up, has given to the opinions of the British Architect a weight and respectability of character, that yields to none of his professional brethren of continental note. The impetus given to that branch of architectural science called Civil Engineering, by the vast and numerous projects for the purpose of railroad conveyance, has called into operation a further development of those energies which have previously been so successfully employed in the art of Bridge Building, and, which called forth the following elegant encomium upon native talent and individual enterprise, from a celebrated French engineer, who, (alluding to Waterloo Bridge) in a memoir addressed to the French Institute, thus apostrophizes his subject :—

"If from the incalculable effect of the revolutions which empires undergo, the nations of a future age should demand one day what was formerly the new Sidon, and what has become of the Tyre of the West, which covered with her vessels every sea? The most of the edifices devoured by a destructive climate will no longer exist to answer the curiosity of man by the voice of monuments, but the Waterloo Bridge, built in the centre of the commercial world, will exist, to tell the most remote generations: here was a rich, industrious, and powerful city! The traveller, on beholding this superb monument, will suppose, that some great prince wished, by many years of labour, to consecrate for ever the glory of his life by this imposing structure. But if tradition instruct the traveller that six years sufficed for the undertaking and finishing of this work—if he learns that an association of a number of private individuals was rich enough to defray the expense of this colossal monument, worthy of Sesostris and the Cæsars, he will admire still more the nation in which similar undertakings could be the fruit of the efforts of a few obscure individuals lost in the crowd of industrious citizens."

It is not, however, the skill or ability displayed by the architect in the structure alone, that enables us to judge of his merits as a practical builder, or that evinces his judgment as an engineer. The attributes of foresight and prudence, so eminently qualified to command success in every relation of life, is, in the case of the architect, a matter of constant and particular application, and upon the due observance of which depends his fame and character as an artist, and his success as a skilful practitioner. To the many, the scientific labours of the architect are but little known and less understood. Buried beneath the bed of rapid rivers, surrounded by the waves of the ocean, or deeply seated upon the firm consolidated stratum of its parent bed, the foundations of his works—the practical application of the study of years—lay concealed from the view of the thousands that admire or impugn the taste of him whose labours are supposed to end with the production of a tasty façade from the drawing board.

The important points involved in the consideration of the present subject has induced the Author to devote this chapter to some general observations concerning the nature of stratification, and, above all, to point out to the inquiring student the necessity of his considering the study of the science of Geology, as embracing a most valuable and significant acquisition to his stock of scientific information. Geology (says the Rev. W. D. Conybeare *) "being the knowledge of the earth's structure, as far as it lies open to our observation, the fundamental point on which it rests, is the ascertaining the order in which the materials constituting the surface of our planet (far beyond this our observation cannot penetrate) are disposed. The superficial and hasty observer might suppose that these mate-

* Geology of England and Wales.

rials are scattered irregularly over the surface, and thrown confusedly together; but a slight degree of attention will prove that such a conclusion would be entirely erroneous." The continued change of locality to which man in his less civilised state was subject, must undoubtedly have led to some particular observations relative to the soil that was to yield him its produce for his support. But the extension of his observations, for any further purpose on this point, may be readily doubted, when we consider the migratory character of his life, and the temporary structure of the rude abode that was formed, rather for the purposes of present shelter and repose than for those domesticated comforts and conveniencies that mark the progressive stages of his civilisation and refinement. Of the progress of architecture as a science, and of the localities of the most antient cities on record, we have given some account in the Introduction; it is now necessary to direct attention to a comparatively modern branch of study, namely, Geology; and to show in what manner the truths unfolded by that science may be rendered of the greatest available service to the student in architecture, whether civil or military, and its professors.

The researches of the geologist having placed before us certain facts relative to the formation of the strata of our globe, it will be necessary, in treating on this subject, to adopt to a certain extent that methodical arrangement exhibited in their most approved treatises. But while we shall have occasion to recur to geological formations in treating on the varieties of building stone, the discussions on the natural phenomena connected with our investigations embraces such an extensive field of inquiry, and is withal so peculiarly the province of the scientific geologist, that a reference * on the part of the author to those standard productions that are most likely to reward the researches of the student, must be considered sufficient for this purpose. We profess only to point out the *application* of this study to the purposes of architectural science.

The distribution of the various strata that form the crust of the earth, are by an intelligent cause so placed before the observation of man, as to point out readily the peculiar characteristics of a country suitable either for the establishment of a city, the occupation of the agricultarist, or the site of the cyclopean works of the workers in metal.

The manner in which the various strata are successively presented to view on the *surface* of the earth is by what is termed (geologically) the "outcrop, or basset of the strata;" this remarkable appearance originates in consequence of the different strata successively overlapping and emerging from beneath each other, † thereby forming a series of

* Phillips and Conybeare, "Geology of England and Wales." In the elegantly written introduction to that work, numerous references are given of considèrable importance, and to these I must refer the student who wishes to extend his geological inquiries, only adding to them the late celebrated "Bridgewater Treatise" of Professor Buckland.

† This natural arrangement of the strata enables us to become acquainted

zones or belts, marking the variations of soil, and distinguishing in many instances one county from another. From whatever point we proceed, considering (for example) London as the centre, the same phenomena and the same distribution of strata exist—subject, however, to occasional interruptions, dislocations, &c., arising from the external agency of atmospherical influences, irruptions of water, or of the no less violent agency of volcanic fires. The operations of the architect, in his search after a solid substratum, do not extend to any very remote depth; for should the soil prove treacherous, he has other and artificial aids for remedying a defect which would probably otherwise require extensive excavations. Previous to entering upon a description of these artificial methods, it will be requisite to point out some of the more prominent strata that will be most likely to command the attention of the young architect or engineer. Geologists have generally agreed to the following classification of the strata as they appear on the surface of the earth:—The *alluvial*, the *diluvial*, the *regular strata*, consisting of the various beds or layers of *sand*, *clay*, and *limestone*, each of which, with their separate layers, are also designated by the comprehensive terms of siliceous, agrillaceous, and calcareous masses, the most prominent of which in this island are the *chalk* and the *limestone*.

with the nature and properties of the beds or layers, otherwise inaccessible from their extreme depth in a vertical direction.

The strata called *alluvial*, is that peculiar deposit which is formed under our immediate observation, and consists of accumulations of sand and shingle along the sea-coast, in estuaries, the alluvial depositions forming new lands on the banks of rivers, lakes, &c., and other apparently minor causes, but which in process of time gradually operate and effectually change the face of well-known localities. The immense quantities of shingle or beach which is now, and has, indeed, for some time past, been collecting together at the foot of the town and port of Dover, is an example of the formation of this kind of deposit worthy the attention of the engineer.*

The *diluvial* deposit is that extensive coating or layer of sand and gravel indifferently covering all the solid strata, and appears to warrant the opinion of Geologists, that the derivation of this universal strata may be traced to one great convulsion which has partly destroyed or broken up the solid strata; this opinion is further verified, or borne out, in consequence of the gravelly portion of the deposit being in all cases composed of debris of that particular rock, most abundant or peculiar to the site. In addition to this, Mr. Conybeare observes,

* The Harbour of Dover is frequently impassable, in consequence of a bar of shingle being formed across the mouth during the prevalence of south-westerly winds. The immense expense attendant upon the extensive alterations of the works, and the question that arises as to their ultimate efficacy, would lead us to infer that the engineers had overlooked that slow, but certain, operating cause that appears to mock their endeavours, namely, the natural formation of an *alluvial deposit*, on this site.

" that between these accumulations of fragmented rocks and the vallies traversing the present surface of the earth, there clearly exists a close relation; that, namely, between the breaches that have been opened in the ruined strata, and the materials which have been removed from those breaches. The same causes that have excavated the one have heaped up the other; and these causes have evidently (as appears from a general examination of the phenomena) acted at once on all the strata, and at a period subsequent to their original formation and consolidation: hence they must be assigned to the last violent and general catastrophe which the earth's surface has undergone, whatever has occured since being either the great action of causes still continuing to operate, or convulsions violent indeed, but of very limited and local extent."

The *regular strata* forms the next general division, and is composed of sand, clay, marle, and limestone, &c., the interminable number of these strata alternating with each other in the order already represented, would prove a source of infinite perplexity to the geologist as well as the student, were not the general arrangements of the former to comprise the numerous alternating strata of one kind or genus under one comprehensive head, thus: the alternating beds of *sand* and *clay*, or *chalk* and *flint,* are, however numerous, referable to the general heads of those two great divisions. The great chalk formation, has been with propriety chosen as the leading feature for the consideration of the geological character of this country, its peculiarity of situation, and the guide that it forms to the observation of other phenomena connected with geological inquiries, has justly rendered this conspicuous strata of first-rate importance in these and similar investigations. One of the most important features of the chalk strata, is that scarcely perceptible dipping under all the other strata before mentioned—forming, when viewed as one great and comprehensive whole, an extensive hollow or basin,* interrupted only by the line of coast: this formation, consequently, becomes the highest of the English strata, as will appear by referring to what has been already stated when speaking of the "outcrop or basset" of the strata. We have hitherto considered the nature of the soil or strata geologically, and in an extended and general sense. The architect will have to judge of the practical value of his substratum for building purposes, by examining the soil more in detail; and, indeed, this portion of his labors is arduous and exciting in the extreme: his excavating operations frequently extending through numerous subordinate strata, in most cases imperfectly consolidated. Other instances occur in which the depth of one stratum, (the London-clay, for instance) is by far too great a thickness to allow

* The most northerly of these basins, including the metropolis, is on that account called the London Basin.

of its penetration to its entire depth. In order to examine the nature of soils for the purpose of building upon; or, for the supply of water, recourse has been had to boring the earth to a considerable depth, thereby enabling us to examine what is brought to the surface by the shell of the boring-rod; this method, however, is far too vague and indefinite to be *confidently* relied upon, for the building of foundations. Its utility in other respects cannot be questioned, and the apparatus itself merits a description in this place, as one of the means, the uses and appliances of which should be distinctly understood by the architect and engineer. The following invention, together with the drawings and description, have been kindly furnished by my esteemed and ingenious friend, Luke Hebert, Esq., C. E.

The implements made use of are extremely simple, and will, I trust, be understood, by reference to the above drawings:—

A, is the cross handle of the borer for two men to work.

B, the chisel borer, which is made to screw into A.

C, the auger which also screws into A.

D, a lengthening rod, screws also into A, at one end, and at the other end it has a hollow screw, into which is either fitted the chisel or auger, or another piece of rod. A great number of these lengths of rod are kept generally in readiness, which are screwed one into the other, so as to proceed to the depth of several hundred feet.

E, a forked iron, used to lay across the hole to support the rods at the joints, whilst the pieces are being screwed and unscrewed.

F, a spanner, used to screw on and unscrew the various tools and lengths of rod.

G, a clearing chisel, with a probe or piercer attached to guide it.

H, the spring bar used to produce a vibrating, up-and-down motion to the chisel, when used to peck away hard or rocky ground.

I, iron chain, to connect the cross handle of the tools to the spring bar.

J, two men at work, boring with the chisel.

K, the lower pulley of a pair of blocks suspended to compasses above.

L, the compasses.

M, winch or crane to work the blocks, when great weights are to be raised.

O, three lengths of rods, and the chisel in the act of boring—perforation, about 42 feet.

P, spiral worm or auger.

Q, square iron bar passing through the square tube of the spiral worm.

R, chains to draw up the spiral worm along the bar.

S, top plate of spiral worm, to which the chains are affixed.

T, upper view of plate, showing the square hole through which the bar passes.

U, angular point of square bar.

V V, cutting auger edges.

W,-underside view of the bottom, cutting parts of the spiral auger.

As a preparatory measure, a large circular hole is usually dug to the depth of seven or eight feet, at the bottom of which a floor is formed by means of some planks for the men to work, and pace round upon, whilst using the implements. * If the earth is very soft, the only tool requisite is the auger C, of three to four inches diameter, which is screwed into the cross handle A, and the perforation is easily effected by the mere turning of it round by two men, as shown in the drawing. When the auger has penetrated to nearly the depth of the tube, it is withdrawn and cleared of its contents; it is then let down again, and the perforation continued to the length of the instrument. To proceed to a greater depth, the lengthening rods before described are put into requisition, the auger is detached from the handle by unscrewing it, a piece of rod D, is screwed in its place, and the auger screwed on to the rod. With the instrument thus lengthened seven or eight feet, the boring is renewed by means of the auger as long as the earth is found to be sufficiently

* Instead of this, sometimes a stage is erected, from ten to fourteen feet above the ground, where the men turn the boring implements, assisted by a man or two underneath it.

soft and yielding: whenever it proves otherwise, or hard and rocky, the auger is detached from the rod, and the chisel B, which is from three to four inches in diameter at its edges, is screwed on in its place. If the ground is not very hard, the boring may be continued with the chisel, by the workmen pressing upon it as they turn it round; but when the earth is too hard to be operated upon by the chisel in this way, recourse is had to pecking, which is done by lifting up the implement and striking it against the opposing substance till it is chipped away, or reduced to powder, to a certain depth. The rod and chisel are then again drawn up, and the auger substituted for the chisel, for the purpose of extracting the pulverised stony matter contained in the hole. The chisel and the auger are thus employed alternately where the ground is hard and stony—the one for chipping away, or pulverising, and the other for clearing out.

As the perforation deepens, the process of pecking becomes very laborious, recourse is therefore had to a very simple contrivance, called the spring-bar, (see H,) which affords the most effectual aid. This is a strong pole, placed horizontally over the well, at the height of three or four feet from the ground, with one end inserted into a spot, or other strong hold. The chain I, is attached to this bar, and the handle of the borer is suspended to the hook of the chain, which supports its weight; a slight vibrating motion is then given to the bar by the workmen, which causes the chisel to peck away with great rapidity.

As the weight of the implements becomes too great to be drawn up by hand, when the boring has proceeded to a great depth, the mechanical aid of a pair of pulley blocks, K, is used for the purpose, which are usually suspended to a tripedal standard, (or triangle as it is called,) fixed over the hole. In withdrawing the rods for the purpose of bringing up the materials bored through, or for changing the tools, every piece is successively unscrewed, and upon re-introducing it into the hole, every piece is necessarily screwed on again, which renders the operation exceedingly tedious. This inconvenience, it appears to me, might be greatly lessened, by erecting a more lofty standard, such as is used in shifting the masts of ships, called compasses, which are formed of two long mast spars, connected together at top with ropes, passing from the summit in a cross direction to the ground (so as to form a quadrangular figure at the base) to secure the compasses in their position. A crane, or simple winch and ratchet wheel, might then be fixed to one of the legs of the compasses at M, which should work tackle blocks, or a single wheel and axle suspended to the upper part of it; which would enable the workmen to raise a great length of rod safely, without the necessity of the almost incessant screwing and unscrewing, which occupies full three-fourths the time and labour. Considering the compasses as preferable to the tripedal standard, I have introduced them, in lieu of it, into my drawing ; see L.

In the manner described the boring proceeds, changing the tools to such as may be best suited to cut through the various strata ; whether of a soft, indurated, or stony texture, until the main spring is arrived at, when the water flows up the newly formed tube to the height of the distant spring from which it is derived. If that be at a greater altitude than the surface of the earth bored, the water rises above the ground, producing a perpetual fountain; on the contrary, if it be below the surface, a well must be sunk, of some capacity, lower down than the level of the adjacent spring, into which the water will flow and form a reservoir, to be drawn up by means of a pump.

The earth is sometimes bored by the beforementioned simple instrument, to the depth of two, three, or four hundred feet, either for the purpose of obtaining water, or to ascertain the presence of minerals. To carry on the operation of these immense depths, it is of course necessary to employ a greater power than that of the two men in the drawing; but any degree of force may evidently be obtained by lengthening the cross bar or levers, and working them above ground, as a capstan on board a ship ; or a horse may be employed to turn the boring shaft the same as in a mill. These contrivances, however, are only my own suggestions, and must be obvious to every body.

When the hole is bored, a pipe of cast iron, or

other metal, is forced down, to prevent its being filled up again by the falling in of the surrounding earth, and likewise to keep out the impure land springs, which might taint the water.

Reflecting upon the excessive labour and tediousness of the ordinary method of boring the earth, I was led to consider of some means by which the operation might be carried forward at great depths, and the earth be extracted without the necessity of withdrawing the rod, and thus save full nine-tenths of the time and labour, which are occupied in the almost perpetual screwing and unscrewing of the various pieces composing its length every time it is let down or drawn up. The method by which I propose this desirable object is extremely simple, and I will endeavour to describe it :—

An auger is to be made with a spiral worm, winding round a cylinder, which is to form its centre.* The cylindrical part is not to be solid, but to be perforated throughout its whole length with a square hole of two inches or more diameter, for the purpose of receiving within it an iron bar of the same figure and admeasurement. The bar will thus serve the double purpose of a spindle, or shaft, to work the auger, and cause it to bore; and as a slide, upon which it may be drawn up with facility from very great depths to the surface in a few seconds of time,

* This figure may perhaps be better understood by comparing it to a circular staircase winding round, and supported by a column or newell in its centre.

contents be discharged, and let down again, as quickly to proceed in the perforation of a fresh portion of earth.

It is perhaps worthy of notice, that an auger with a spiral worm is, independently of the other circumstances mentioned, much better adapted to

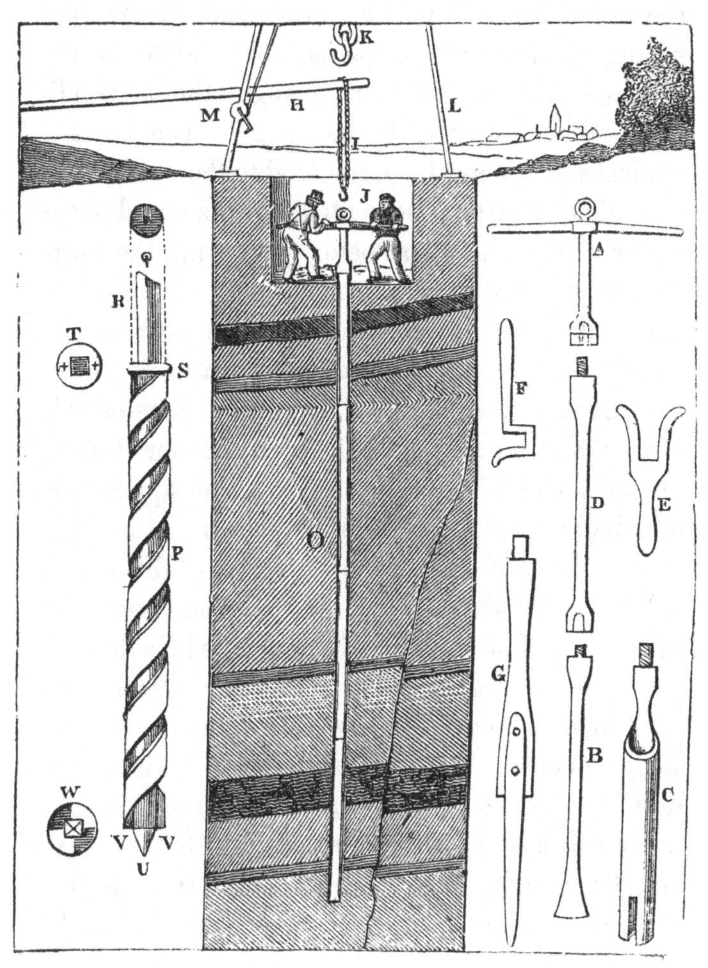

boring the earth, than the common auger. First, because it requires less power to force the earth up the inclined plane of the spiral auger, than perpendicularly up the common auger; consequently, the latter, by the application of an equal force, cuts its way more slowly. Secondly, because the weight of the earthy contents in the spiral auger, lying supported on its inclined plane, adds force to the cutting, while in the common auger the perpendicular column of earth in its centre has a bearing against the edge of it, prejudicial to the cutting.

That a part of the contents of the spiral auger may not fall out, when being drawn up, the worm or thread is not to be left open, but to have a perpendicular border, raised upwards at right angles with the plane of the thread, as shewn in the drawing; the aperture between the upper edge of this border, and the next thread, is left open for clearing out the auger with facility, see P. The square bar, or spindle of the auger, is shewn at Q. R are the chains attached to the top plate S of the auger, and passing over a pulley. T shews an upper view of the plate of the auger, with the square hole through it. U is the angular point of the bar, which may be formed as a chisel or any other figure. V V are the side cutting edges of the auger. W is an underside view of the auger, shewing also the two under edges which are connected with those at V V, and form right angles with them. It is evident that various kinds of tools may be fixed at the bottom of this auger so as to peck, &c.; but to go into particulars on these minor points will perhaps be tedious, and extend this description, already, I fear, too lengthy. I must not, however, close it without remarking, that the auger may be easily latched or bolted in its situation, should it be considered requisite, and unfastened on pulling the chains tight; but I am of opinion, that the weight of the instrument would render the fastening of it down unnecessary, as any required force might be given by loading the upper part." The various figures referred to are fully exhibited in plate 1.

Considerable ingenuity has been displayed from time to time in the construction of the various implements used for boring the earth; among these, honorable mention may be made of those invented by Mr. John Good, of Knightsbridge—the *modus operandi* being the same as that already so ably described by Mr. Hebert; I shall merely give, in the words of the same writer, the following description of the instruments invented by Mr. Good, referring to plate 2 for the illustrative drawings.

It will be noticed that all the instruments, except four, have a screw at their upper ends, which are tapped to one thread, so as to fit uniformly into the rod or rods by which the process of boring is conducted. Every rod is screwed in like manner, so that any number of rods may be connected by their screws, end to end, and on the lowermost of the series of rods is fixed the boring implement.

Fig. 1 represents the principal instrument employed for boring through soft earth, such as clay, loam, chalk, &c. : it is of a cylindrical form, with an opening or slit down its whole length, equal in width to one third of the circumference of the cylinder ; this construction having been found by the patentee to be the best suited for earths of a compact but not very hard nature. When, however, the earth is very soft and yielding, an auger with a narrower slit is provided, only proportioning that part to the degree of tenacity of the earth which is to be extracted. At the bottom of this auger a cutting edge is fixed at *a*, by screws, and adapted so as to take off and on as may be required, either to change or sharpen the instrument.

Fig. 2 is an implement of a hollow conical form, with a spiral worm wound round it ; which is for the purpose of boring through very loose sandy soils ; the sand, as the boring proceeds, passing along the inclined plane of the screw, until it arrives at *b*, when it is discharged into the hollow receptacle of the core.

Fig. 3 represents another form of auger, which is employed for extracting mud, sand, and other very soft matter ; it is a close cylinder, with a valve at *c*, part of the cylinder being broken away to bring it into view. It has a cutting edge, A, at bottom, as Fig. 1, and is employed in like manner : upon turning this auger, the soft, yielding, or fluid matter passes through the valve and fills the cylinder, which, on being drawn up, the valve closes, and secures the conveyance of the contents.

Fig. 4 is a close cylindrical vessel, of the nature of a bucket with a valve at the bottom (as shewn). Having a rope fixed to the upper end, it is dropped down the hole, and, being alternately raised and let fall, it acts as a pump, to extract the liquid or floating matter that may be contained in the hole. Fig. 5 is another pumping tool, more complete than the former; the cylinder being fitted up in the manner of the common lifting pump with a rod and bucket, *e*. In this instrument the rod is lifted at every stroke, instead of the whole apparatus, as in fig. 4, and has the preference more particularly in great depths. There are two stops and a guide piece at *f*, to limit the extent of the strokes.

Fig. 6 is an instrument for extracting rods when broken in the hole: in its upper part it is like a pair of tongs, having at the lower end a cylindrical tube, and above it, at *g*, a chisel-edged tongue, which is pressed downwards by a spring. Upon lowering this instrument into the hole, the upper end of the broken rod passes through the tube, pressing back the tongue, which holds it fast and prevents its returning, so that it may then be drawn up out of the hole. This last mentioned form of instrument is sometimes employed with two pressing edges or tongues, similar to those shewn in fig. 7, at *h*, which is employed for precisely the same purpose; the operation and construction of which will be best understood by referring to the engraved representation. Fig. 8 is another tool for raising broken rods, and adapted so as to lay hold of the rods by means of its chain underneath their joints, and by that means be drawn up.

Fig. 9, is an improved chisel-punch, employed for perforating stony matter by repeated pecking; the projecting middle piece, which strikes first, the patentee finds to be of great advantage; the blow is more effectual, and the fractured stone is thereby more easily displaced. Fig. 10 is another form of chisel, for chipping away hard substances, and by its peculiar construction likewise adapted to the operation of boring; the sides of the instrument are bent in an angular form, so as to present cutting edges, rounding and perfecting the whole as it is turned by the workmen. Its peculiar form is shewn by the edge view, Fig. 11, *i* Fig. 10, and *i* Fig. 11, representing the same parts.

Fig. 12 is the apparatus employed for ramming down the iron or other metal tubes which line the perforation after being completed. A block of wood, *k*, has an iron rod, *l*, fixed perpendicularly in it; on this rod a weight, *m*, is made to travel or slide by a hole being passed through it; the wooden block being fitted into and over the edge of the upper end of the metal tube, *n*, the weight is raised by cords, and let to fall upon the wooden block, thereby forcing down the pipe. When a piece of tube has thus been driven down even with the surface of the earth, shewn at *o*, another piece is fitted into it, and the operation of ramming down renewed as before. Another instrument for this purpose is sometimes employed, of the shape of an acorn, shewn at Fig. 20, which is made to screw into the end of the perforating rods, and being let down into the hole, the lower part of the acorn tool enters the pipe, and the projecting rim resting upon the edge of the tube, admits of its being forced down with an even bearing pressure. Fig. 13 is an instrument constructed like a pair of tongs, armed at each end with two projecting cutting edges, as at *p*, which, on being turned round, cut like the sides of an auger or gimblet, and are thus employed in paring or widening the perforation. Fig. 14 is a tool formed like a double bow, employed for pressing out any indentations that may be accidentally made in the lining tubes, which are sometimes made of a soft metal, such as lead, or of a thin substance, such as copper or tinned sheets. Fig. 15 is another tool for the same purpose, forming a quadruple bow; it is adapted for rubbing down smaller indentations, or to finish the work of the double bow tool. They are both fixed like the other tools by screwing into the end of the rod, or series of rods, and operate by simply turning them round.

Fig. 16 is a pair of circular clans employed for holding or turning round the tubes when being sunk; being fastened together by means of screws, the pipe may be firmly held within their grasp.

Fig. 17 is a tool of a pine-apple shape, used for drawing up a piece of tubing when required. It is jagged all over like a rasp, and being struck down forcibly within the tube, it takes fast hold, and allows of its being drawn up. Fig. 19 is another implement for effecting the same object; it is a kind of spear with four prongs: being forced down, the points pierce the tube, and enables it to be withdrawn.

Fig. 18, is a triangular instrument with jagged teeth, used for getting up loose stones which sometimes lie in the hole and impede the work: upon being struck down, the stone becomes fixed between the notches of the prongs, and is then extracted.

At twelve or more feet from the ground, a stage is erected over the hole. Upon this stage is fixed a double-handed winch, with tackle complete, for two or more to work, in raising or lowering the instrument, &c., which becomes of great weight when the perforation has proceeded to a considerable depth. Between this stage on the ground the men are employed either in turning round the instruments in the act of boring (which they do by the help of iron levers fastened crosswise to the boring rods by means of screw bolts), in screwing or unscrewing the rods and tools, or in the various other operations before described.

The examination of the strata may be conducted in various ways. The geologist observes the position and quality of the strata, as shewn on the surface of the earth, by its "outcrop or basset," by the natural sections laid open to his observations in the cliffs of the sea shore, or by descending mines, quarries, and pits. The architect, it is true, has also these sources of observation at his command; but his investigations must be of a more searching character, and being confined to a certain spot, his proceedings must be modified according to the exigencies of the case. These are neither few nor altogether easy of attainment, for although we have observed (p. 6) that "the operations of the architect, in his search after a solid substratum, do not extend to any very remote depth," it must be understood, that the observation is applied as a comparison between the excavations necessary for the erection of an extensive edifice, and the more magnificent operations of the same kind that are effected in the mining districts of Europe.* What-

* The following are the depths of the deepest shafts in the world:—

	Fathoms.	Feet.
1. The shaft called Rochrobichel at the Kitspüel in the Tyrol	460	2,760
2 A Copper Mine, not now at work, at the Southampton Silver Mine at Andreasburg, in the Hartz	371	2,226
3. Valenciana Silver Mine at Guanaxuato, Mexico . . . Began in 1791, and reached its depth in 1809, when it was stopped by the Revolution; it is probably the finest in the world, being octangular 30 feet in diameter, and a great part walled with masonry; expense of sinking estimated by Humboldt at £220,000.	295	1,770
4. Pearce's Shaft at the Consolidated Mines (Copper), Cornwall	244	1,464

ever may bethe depth for the proposed foundation-levels of a building, the scrutiny of the architect or the engineer should in no wise rest there; the strata that underlies the intended works, is not unfrequently discovered to be of such nature as to subside gradually as the weight is added, carrying with it the portion of the building that stands thereon, together with its artificial foundation. Hence the expedient of boring the earth to ascertain the nature and number of the different strata, and, as far as is practicable by such means, to decide upon the quantity of earth to be excavated, and the length of piling required, should that method be adopted for the construction of the underworks. This preliminary procedure in practical operations is the more essential, as it places us in possession of

	Fathoms.	Feet.
Not sufficiently productive to repay the draining.		
5. Whieal Abraham Copper Mine, Cornwall	242	1,452
6. Dolcoath Copper Mine, Cornwall	235	1,410
Produces nearly half the amount of the Consolidated.		
7. Ecton Copper Mine, Staffordshire	230	1,380
Once one of the principal Mines in England, but now inconsiderable.		
8. Woelf's Shaft at the Consolidated	225	1,350
The bottom of this shaft is nearer the centre of the earth than any other known point.		
The deepest coal Pit near Newcastle is Innou Pit, near South Shields, which is	165	990
Working below the bottom of the pit	12½	75
Total . .	177½	1,065

certain facts and data, relative to the nature of the subsoil, without which the necessary examination of the foundation-levels must have been both tedious and expensive. The excavations necessary for building purposes being generally commenced from the surface, and carried on in a vertical direction downwards, we must be allowed to consider the chalk stratum *as the lowest*, in contradistinction to the geological arrangement that places it as the *highest stratum*, when viewed with reference to its outcropping position. The strata immediately resting upon the chalk basin of London, consists of a series of layers, loose and friable in their textures, namely, white sand, pebbles, marl, and shells; upon this succeeds the "plastic clay formation," and above this the great formation, called the "London clay." This latter strata is in some parts immediately under the vegetable soil, and forms the general substratum of the metropolis; hence its name. As there are certain exceptions respecting the depth at which this strata first appears, we may observe, that occasionally it is covered with the general diluvial deposit to some considerable depth. The following remarkable instance of the position of the London clay was exhibited in preparing for the reception of the foundations of St. Paul's Cathedral. The circumstance being barely mentioned by some, and incorrectly described by other writers, the author submits the following extract from the "Parentalia," as the most authentic description of the proceedings of the architect (Sir C. Wren), in his search after a solid substratum.* "In the progress of the works of the foundations, the surveyor met with one unexpected difficulty: he began to lay the foundations from the west end, and had proceeded successfully through the dome, to the east end, where the brick-earth bottom was very good; but as he went on to the north-east corner, which was the last, and where nothing was expected to intercept, he fell, in prosecuting the design, upon a pit, where all the pot earth had been *robbed* † by the potters of old time. It was no little perplexity to fall into this pit at last; he wanted but six or seven feet to complete the design, and this fell in the very angle north-east. He knew, very well, that under the layer of pot-earth there was no other good ground till he came to the low-water mark of the Thames, at least forty feet lower; his artificers proposed to him to pile, which he refused: for though piles may last for ever, when always in water (otherwise London Bridge ‡ would fall), yet, if they are driven through dry sand, though sometimes moist, they will not. His endeavours were to build for eternity. He, therefore, sunk a pit of about eighteen feet square, wharfing up the sand with timber, till he came, forty feet lower, into water

* "Parentalia; or, Memoirs of the Family of Wrens," by Stephen Wren, Son of Sir Christopher Wren; folio, edit. 1750, p. 286.

† Taken from, or dug out.—C. D.

‡ In allusion to the old structure, now removed—the timber of the piles and the iron shoes at their extremity, were found sound.—C. D.

and sea-shells, when there was a firm sea-beach; which confirmed what was before asserted, that the sea had been, in ages past, where St. Paul's now is. He bored through this beach till he came to the *original clay*; being then satisfied, he began from the beach a square pier of good solid masonry, ten feet square, till he came within fifteen feet of the present ground; then he turned a short arch underground to the former foundation, which was broken off by the untoward accident of the pit." The number of extensive works with which the metropolis abounds, and the number of wells, of almost every description of depth, that have been dug, present us with a mass of information connected with the sections of the strata immediately above the great *chalk formation*, that is of signal service to the architect and engineer, should his operations occur in the locality of one of these chalk basins. The extreme density of the *London clay* is shewn to afford a highly favourable substratum for the erection of heavy works, where it is not intersected by the river or streams. The plastic clay formation, on the contrary, is exhibited as containing numerous springs, and is altogether of such a nature as to warrant its abandonment as an efficient strata for the erection of foundation walls, &c. &c.

The greater part of the London clay strata is of one uniform colour, extreme density, and variable thickness. It is so impervious to water, that that necessary article cannot often be obtained until after the stratum has been perforated to the depth of the plastic clay formation. The inhabitants of London and its vicinity have, however, a resource which forms one of the greatest manufacturing coveniences of the metropolis. This consists in the immense quantities of water held by the alluvium, forming supplies to many of our factories, without the intervention of land springs, at a greater depth. Examinations of the strata near London * have been made by several scientific gentlemen. A few observations made on the subject of the *blue clay stratum*, by Mr. Parkinson, in a paper published in the "Transactions of the Geological Society," states the depth to be 200 feet in thickness ;† its colour for a few feet in the upper part was observed to be of a yellowish brown, but the remaining portion exhibited that dark bluish grey colour which marks its genera. It has been suggested, that the difference of colour remarked between its superior and inferior part arises from a difference in the degree of oxidation of the iron present in it: the result of a more particular examination would appear to justify the supposition, that it arises rather from a difference of *quantity*, than the degree of oxidation mentioned above; for "it is probably occasioned," observes Mr. Parkinson, "by the washing away of this metal in the water which percolates

* Transactions of the Geological Society, vol. i. p. 336. "Mr. Parkinson on the Strata and Fossil Remains near London."

† The particular locality is not mentioned. At Epping, near High Beach, it is about 700 feet in thickness.—See Conybeare and Phillips.

through it, and which runs off laterally by the numerous drains made near the surface." The following sections, afforded by the preparation for the progress of several public works, will further exhibit the nature and uniformity of the London strata, and will also serve as a guide to the examination of other strata situated within or near the localities of the several chalk basins. The identity of strata on the French coast, opposite Dover, with that of our own chalk formation, will be dwelt upon in treating on the *Manufacture of Bricks.*

SECTIONS OF STRATA

ABOVE THE CHALK BASIN OF LONDON.

Names of Works.	Depths examined.	Feet.	In.		Authorities consulted
New Pancras Church.	1. Gravel, &c. . . .	6			Seabrook, Esq.
	2. Yellow clay, about .	2			
	3. London clay (hard)	82		no water	
Thames Tunnel. South Shaft.	1. London clay	9	0		Conybeare.
	2. Loose watery sand and grav.	26	8		
	3. Blue clay	3	0		
	4. Loam	5	1		
	5. Blue clay with shells, chiefly cytherea	3	9		
	6. Hard conglomerate rock, consisting of flint gravel, with a calcareous cement	7	6		
	7. Light blue laminated clay, with pyrites	4	6		
Boring at Brentford, six miles from London.	Brick earth	9			W. Gravatt, Esq. (See Trans. of Institution of Civil Engineers, vol. i. p. 151.)
	Sandy gravel	7			
	Loam . .	5		varies from 1 to 9 feet	
	Sand & gravel	4		varies from 2 to 8, containing water.	
	Blue clay .	200			
		225		Boring discontinued still in clay.	

Names of Works.	Depths examined.	Authorities consulted.
Primrose Hill Tunnel for Birmingham Railway.	This Tunnel passes through the plastic clay formation. The water percolates rapidly through the brickwork near the entrance (Feb. 1827) ; but the remaining part of the Tunnel is dry and well ventilated.	C. D.
Particulars of a well sunk at the Excise-office in Broad-street, London.	In the first place, after excavating the upper stratum of gravel and loose soil, four cast-iron curbs were sunk, each six feet long; the lowest of these entered the clay about 3 feet; the digging was then continued through the clay to the depth of 140 feet, and a curb of brickwork within the iron curb was sunk the whole depth in the ordinary way, the iron curb serving merely to support the upper stratum, and to prevent the land water getting into the well. Boring was then resorted to, to the depth of about 20 feet, when the water appeared, and rose to within 60 feet of the top of the well; a copper pipe was then driven through the last mentioned 20 feet, to keep the passage open for the supply.	John Donkin Esq. (See Trans. of Institution of Civil Engineers, vol. i. p. 155.)
Church of St. Mary-le-Bow, Cheapside. (See Engraving, Plate 3.)	This church (Tower) stands upon an ancient concrete causeway 18 feet below the *present level*, which present level has been artificially raised during the course of some centuries.	Paper by Mr. George Gwilt. (See "Vetusta Monumenta," 1835.)

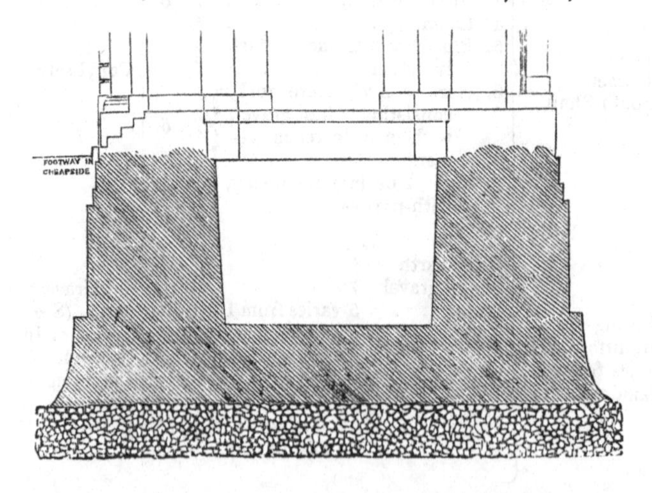

Our observations upon the strata immediately above the chalk need not be extended any further, since from experience and investigation the following facts may be deduced:—1st. That there are only two strata of the series alluded to, which possess the advantage of sufficient consolidation for the support of works of magnitude, these are the gravel and the London clay. As the thickness of all strata varies considerably, some general rule ought to be adopted to secure the permanent stability of the intended structure. Referring to the former strata we may remark, that when there occur a series of thin beds, almost approaching to laminæ, they should be wholly discarded, or such means adopted as are hereafter mentioned, to render the substratum of sufficient density for the intended purpose. But when the bed of gravel is consolidated, or partaking of the nature of a conglomerate, and of a thickness equal to the mean distance between the principal points of support, then the foundation may be deemed secure, and built upon accordingly. The London clay strata is so well known to most metropolitan builders and architects, as to need but slight remark from us. Its extreme density and uniformity of character renders it one of the most secure of the natural strata. The extent of its subsidence by compression is trifling, and its thickness for the most part is very considerable; there are, however, certain situations near the banks of the Thames where this strata exhibits a sudden deficiency of depth, consequently piling has been resorted to in many extensive works in that locality. It will appear obvious that much must be left to the skill and ability of the architect, his proceedings may have to undergo considerable modifications according to the exigencies of the cases as they occur in practice; for, although we may cite precedents for professional guidance, it will be necessary, in the first instance, to point out the difficulties that are most likely to beset the practitioner in the course of his operations, and the means best adapted for the successful accomplishment of the end he has in view.

CHAPTER II.

ON PILE DRIVING AND PILING.

WITHOUT some artificial means for consolidating loose and friable strata, extensive excavations must frequently be made for works, whose extent and magnitude are not otherwise of sufficient importance to warrant so costly a proceeding.

On the other hand, where the removal of extensive tracts of objectionable strata would be desirable, but is prevented by the interference of some collateral cause, then the operation of artificial consolidation must be resorted to, either by piling or concreting the entire surface. The artificial substratum, so frequently met with in the metropolis and its suburbs, technically termed *made earth*, consists of dry rubbish and refuse, with which deep hollows are refilled ; or otherwise, consist of loose earth, thrown up from the excavations of a contiguous work. Mounds of this kind, or the treacherous surfaces of the *ci-devant* hollows, are excessively annoying, and, if built upon, require extreme precautionary measures. The foundations of the celebrated Albion Mills, near Blackfriars, were laid upon a substratum of this kind, and afford a good example of the successful application of scientific resources to a work of considerable magnitude and difficulty. Mr. Farey* thus describes the plan, "The building for the Albion Mills was erected upon a very soft soil, consisting of the *made ground* at the abutment of Blackfriars Bridge; to avoid the danger of settlement in the walls, or the necessity of going to a very unusual depth with the foundations, Mr. Rennie adopted the plan of forming inverted arches upon the ground, over the whole space on which the building was to stand, and for the bottom of the dock. For this purpose the ground upon which all the several walls were to be erected, was rendered as solid as is usual for building, by driving piles where necessary, and then several courses of large flat stones were laid, to form the foundations of the several walls; but to prevent any chance of these foundations being pressed down, in case of the soft earth yielding to the incumbent weight, strong inverted arches were built upon the ground, between the foundation courses of all the walls, so as to cover the whole surface included between the walls; and the abutments or springings of the inverted arches being built solid into the lower courses of the foundations, they could not sink, unless all the ground beneath the arches had yielded to compression, as well as the ground immediately beneath the foundation of the walls. By this method the foundations of all the walls were joined together, so

* Treatise on the Steam Engine, by John Farey, p. 515.

as to form one immense base, which would have been very capable of bearing the required weight, even if the ground had been of the consistency of mud, for the whole building would have floated upon it, as a ship floats in water; and whatever sinking might have taken place, would have affected the whole building equally, so as to have avoided any partial depressions or derangements of the walls; but the ground being made tolerably hard in addition to this expedient of augmenting the bases by inverted arches, the building stood quite firm."

Notwithstanding the considerable density exhibited by the London clay, such density is by no means invariable. The sandy, loose, and muddy layers, that occasionally intervene throughout this formation have been repeatedly met with, and caused considerable annoyance and interruption in the prosecution of works that required extensive excavating operations. The excavations of the Albion Mills, the Highgate Tunnel, the Thames Tunnel, the Catherine Docks, and many other public works in this metropolis, have proved to be so many exceptions to the invariable density of the London clay. It must be evident, therefore, that piling is of necessity resorted to, in those instances, where the thickness of the loose strata is too great to admit of a denuding operation being effected. There is another circumstance that merits particular notice as connected with the foundations of buildings upon the clay. This is, the sudden shelving or thinning of the strata near the banks of the Thames on the Kentish shore. The perforation of a few feet from the surface enables us to pierce through the London clay, and enter the next formation, called the plastic clay. Now the nature of this latter strata is frequently such as to yield considerably with the addition of a moderate weight placed upon its surface. Its depth being considerable, piling is not always advantageous, as, on the one hand, the density of the soil will not admit of sufficient friction on the sides of the piles, and on the other, the depth being too great for the feet or pointed ends of the piles to pierce entirely through and rest upon the chalk. In a case of this kind, therefore, a concretion of the entire surface is, perhaps, the most secure artificial substratum. Buildings may, however, be erected upon a yielding soil when certain means are taken for their proper equilibration, as is shown in the example of the Albion Mills. Piles and piling are of various denominations. The square pile, for heavy works, consists of a long piece (or *stick*, as it is technically termed) of timber roughly squared, bound at the upper extremity with a hoop of iron to prevent its splitting from the blows of the ram, and protected at the lower extremity with iron shoes forged to the shape required, in order that the pile may be driven with greater certainty, and through any obstacle that may obstruct its progress downwards. Previous to entering upon a further investigation of the nature of piling, we shall direct the attention of the reader to the various machines used for the purpose of driving piles into the ground. The old pile machine represented in the annexed figure, is composed of two upright

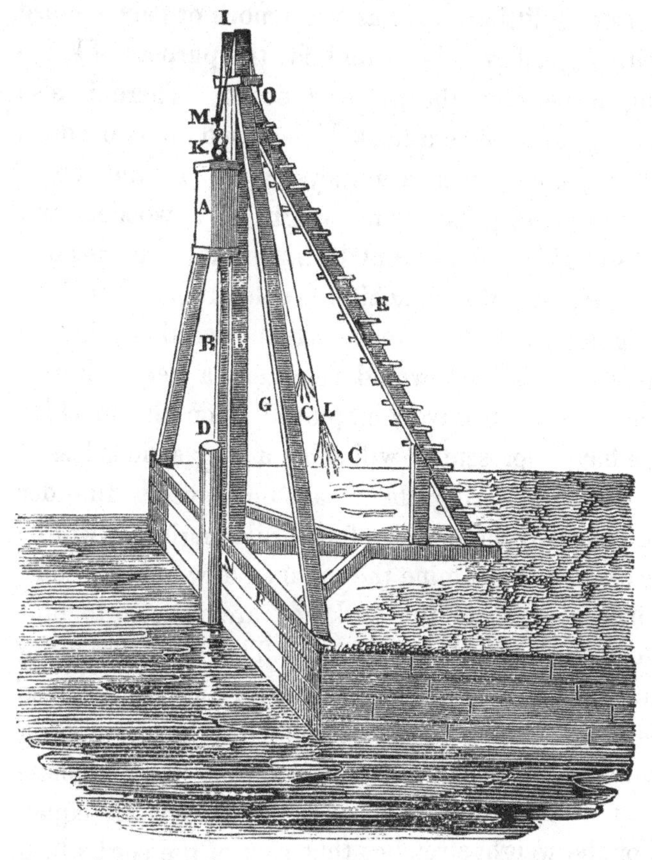

grooved posts B, supported by side braces G G, fixed upon the sole F, and attached to the back stay E; this is connected above to the two uprights, and below to the forked framing H, also secured to the sole F. The ladder for mounting to the top of the machine is shown on the back stay E. The ram A, destined to drive the pile D, is, in this instance, supposed to be a large block of well seasoned wood, firmly bound with iron, top and bottom, to prevent it from splitting. It has two tenons or ears stopped with keys, driven in behind, for the purpose of keeping it between the grooved posts. There is also a ring to receive a hook K fastened to two cords, each passing over a pulley placed at the top of the two uprights. The weight of a wooden ram is generally about 800 lbs., so that the effectual working of this machine would require at least a gang of twenty men, whose exertions, individually considered, would represent a very effective force, but, collectively employed upon the machine we have represented, will show a very serious loss of power, considering the means employed. In order to ascertain the defects of this pile-engine, it will be necessary to examine the manner in which the combined energies of the workmen are directed to produce the action of the ram. The arrangement of this machine is that of a cord passing over a *fixed pulley*. The tension of the cord being uniform throughout its length, it follows, that in an application of this kind the power and weight are equal. For the weight stretches that part of the cord which is between the weight and pulley, and the power stretches that part between the power and the pulley. And since the tension throughout the whole length is the same, the weight must be equal to the power.*

* Lardner.

Theoretically considered, we obtain no mechanical advantage by this use of the pulley, but in its practical applications we are enabled to give to the moving power a more advantageous direction, and with greater facility than without such adventitious aid. Muscular energy may therefore be exerted variously by the means already described. By muscular exertion a man may so far overcome the resistance as to raise his own weight: but in ordinary cases, in raising the arms to lay hold of the cord, and then bending himself as low as is convenient, he can scarcely raise above 70 lbs.; which is about half the weight of his body, commonly estimated at 140 lbs. We have supposed the cord to be drawn vertically; for if it were drawn sideways, the weight or force applied to it will decrease as the obliquity increases. To apply this reasoning to the machine in question, we are to consider that the ten men who are exerting their powers at the ends of the cords attached to the main ropes I L, can only draw obliquely to the vertical line I N, and the same also with respect to the main ropes I L; since in both cases they form a circle, in which those who are opposite to the others destroy reciprocally a part of their power, which is here an inevitable defect: since it is necessary that the ten men should be placed around the machine in order to act together. It is not the same with the obliquity I L; this may be applied in a more vertical direction, by passing the cord M I L over a wheel, 4 or 5 feet in diameter, instead of the pulley, which seldom exceeds 10 or 11 inches. The improvement effected by the use of the wheel experience has fully confirmed; sixteen men being enabled thereby to effect the work of twenty men by the former method. It remains now to be shown in what manner the machine is to be constructed, to supersede the use of the pulley. By referring to figs. 1 and 2 will be seen the plan and side elevation

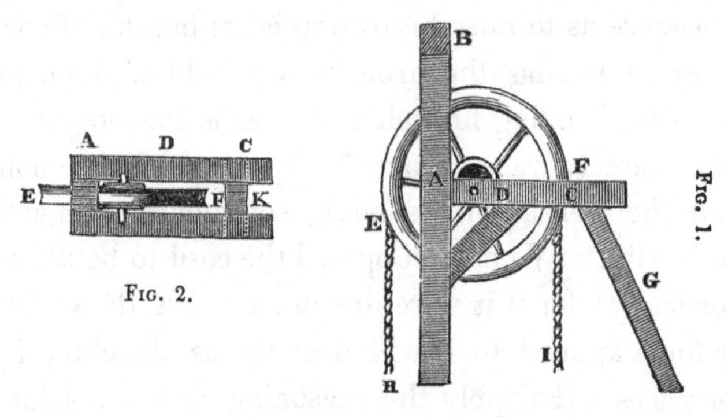

Fig. 2.

Fig. 1.

of the apparatus. To render the pile-machine susceptible of the preceding advantages, it is only necessary to modify the top part of the machine. A marks the two grooved uprights, crowned with a cap B; each of the posts is connected with a piece placed horizontally 3 feet under the cap B; these pieces are fastened together with a cross-quarter of timber, supported by the back stay G, which is fixed in such manner as not to obstruct the position of the wheel E F, which has its nave pierced with a square iron, from which the axle projects, and rests upon its bearings D. More steady action and solidity is gained by this method than if they were not united with the square iron inserted into the nave of the wheel. The wheel is supposed to be 4 feet 6 inches in diameter, measuring from the rope channel, and distant from the posts A about 3 feet. This arrangement enables the men to operate with greater facility, having more room for placing themselves around, and enabling them to pull in a more vertical direction at the other end E H. The more so, as the back stay and ladder G not being in the way, the sixteen men are employed in pulling only one rope, and not divided into two separate gangs—the action in the latter case never being perfectly alike. Another advantage to be derived from the use of the wheel is, that there is only one movement required to counterpoise it, namely, about one third of the circumference of the wheel for each blow—it becoming greater by the acceleration that the ram acquires. The pulley on the contrary, in consequence of the smallness of its diameter, makes nearly two revolutions when the ram rises or falls, thereby causing additional resistance by the stiffness of the cords. In conclusion, we may remark, that, rude as the present contrivance may appear, a pile may be driven by it with great expedition, as it is effected by a series of smart shocks given in succession. The workmen should, however, take advantage of the rebounding of the ram by pulling immediately that effect takes place; much labour may be saved by attention to this point. The advantage of this form in isolated situations is obvious. A workman, unaccustomed to the arrangements of machinery, may in a few hours construct a powerful machine like the one represented. By inclining the grooved upright posts (technically termed *shears*) the machine may be altered for the purpose of driving piles in inclined positions for embankments, abutments, &c. It must at the same time be distinctly understood, that it will not bear to be brought into competition with the machines now in general use, where two men, with the aid of a wheel and pinion, will effectually perform the driving of larger piles that would require sixteen men using the one described, and for which M. Belidor has made his calculations. Figs. 3 and 4 are further improvements on the ordinary machine, and are intended for raising a weight or ram of 12 or 1,500 lbs. by the action of three or four men applied to the windlass A, fig. 3, or the capstan B, fig. 4. It is evident that longer time is required for working this machine, but the force is considerably greater: therefore machines, combining this arrangement, are more generally employed for heavy works, such as bridges, &c. The power, however, is far from being applied in a vertical direction; and time is also lost in releasing the hook C from the weight E, as well as in replacing it after the ram has descended. The advantages of fig. 3 chiefly consist in its capability of being used for driving piles in angles as well as the addition of the hook C for releasing the ram by pulling the rope F, which could not otherwise be accomplished. When piles are driven in rows fig. 4 is more convenient, as it admits of being easily

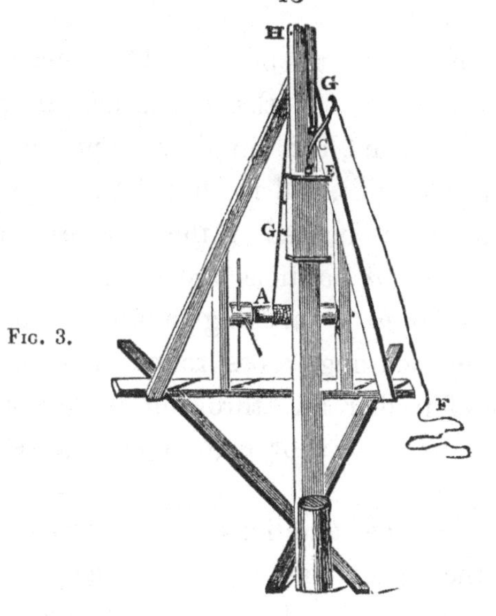

Fig. 3.

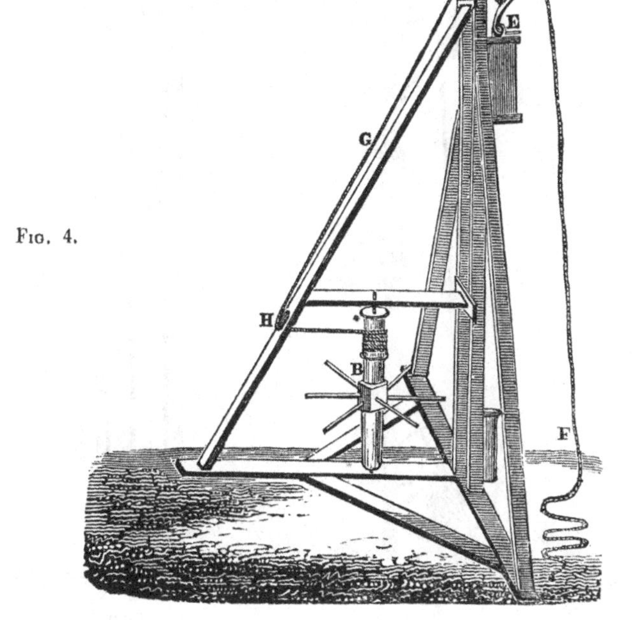

Fig. 4.

removed from place to place. The shears are placed in the centre of the triangular framing. Fig. 3 has no shears, but simply an upright post. The ram slides on this, and is guided by two pieces of bent iron or ears instead of the grooves. Piles driven in horizontal positions are either driven by an engine similar to a battering-ram, or, according to Belidor, by slinging ropes round a long and heavy piece of timber, and throwing it forward by the united power of six or eight men against the head of the pile.

The successive additions and improvements effected in the construction of the pile-engine were the pincers or tongs, under several modifications; followed by the engines of Valoue and Bunce. A

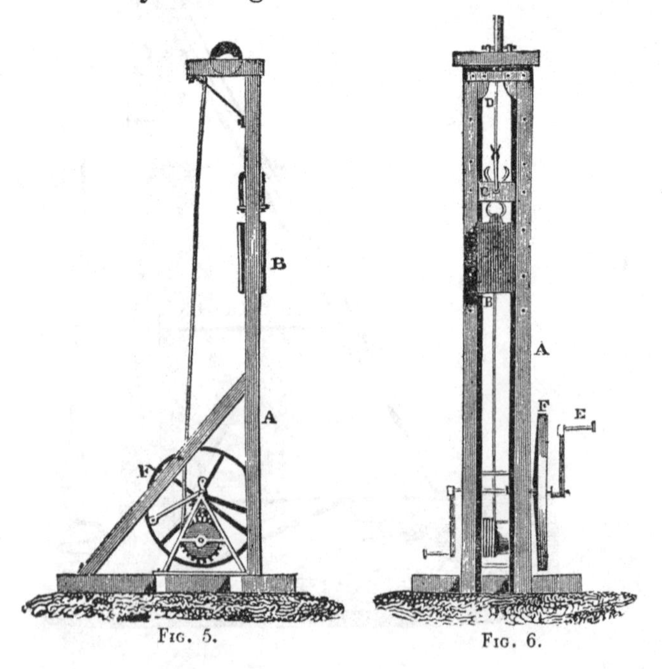

Fig. 5. Fig. 6.

further improvement on the last mentioned machine has left the pile-engine in that state of perfection that we now find it. Its simplicity and efficacy are such, that it has nearly superseded all others, and is the modification now universally employed in large works where piling is required. Figs. 5 and 6 represent the improved pile-engine.

A A are the grooved upright posts, having pieces bolted on, forming the grooves for the cast-iron ram B B to slide in. Two men working at the handles E E cause the drum to revolve, round which the rope coils. The motion of the drum is communicated by a pinion fixed on the axle of the fly wheel F, and takes into a toothed wheel placed on the axle of the drum. On this axle also is a ratchet and pall, to prevent the ram falling should the workmen leave the engine. The pincers A A and D D (figs. 7 and 8) are fixed to a block of

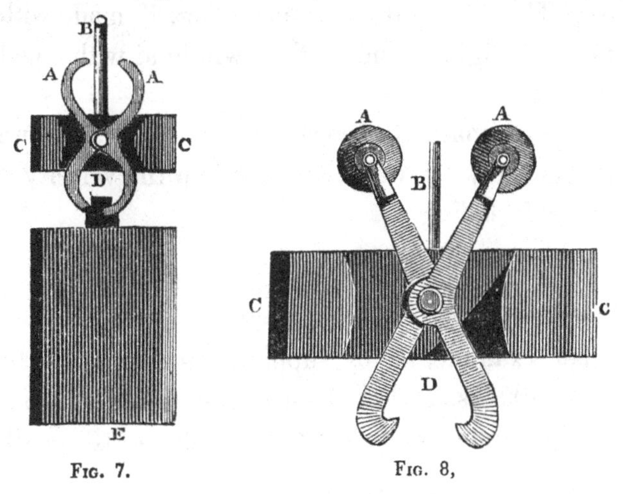

Fig. 7. Fig. 8,

wood, C C, (same as fig. 8) called the *follower*,

sliding in the same grooves with the ram. The rope is fixed to the iron B B (figs. 7 and 8) in the centre of the follower. Now, the ram being raised, the ends of the pincers will pass between the inclined planes D, at the top of the engine (fig. 6), and by compressing the top of the tongs will open the lower ends of them, which clip a ring on the top of the ram, and consequently liberate the weight and allow it to fall on the head of the pile. The pall is then released from the ratchet, and the tongs and follower slide down and again lay hold of the ring on the top of the ram D, (fig. 7.) Occasionally the tongs are made with rollers, A A, (fig. 8) on the ends, in order that they may pass with greater ease between the inclined planes. When the tongs were first introduced, a strong spring was fixed upon one of the levers to keep them closed when they had taken their hold of the ram. This is now discontinued ; for, if made with sufficient weight of metal, they will hold with equal security.

To estimate the power of these machines we must multiply the weight raised into the velocity it acquires in falling.

Let $w.$ = Weight of ram
$v.$ = Velocity in feet per second
Then $w. + v.$ = Momentum of stroke upon the head of the pile

As per example, let us suppose the weight of the ram = 800 lbs., and the height fallen through in one second of time = 16 feet,* then 800 + 16 = 12-

* See Law of Falling Bodies.

800, the force with which the pile is struck by the ram. If pulleys or a windlass be employed to raise the weight to a considerable height and then suddenly disengage it, then the momentum of the stroke will be as the square root of the space fallen through. The inductive reasoning on this subject by an intelligent Engineer (Mr. Nicholson) is thus explained :—" Notwithstanding the momentum or force of a body in motion is as the weight multiplied by the velocity, or simply as its velocity when the weight is given, or constant; yet the effect of the blow will be nearly as the square of that velocity, the effect being the quantity the pile is driven into the ground by the stroke. For the force of the blow, being transferred to the pile, being destroyed in some certain definite time by the friction of the part within the earth, which is nearly a constant quantity, and the spaces in constant forces being as the squares of the velocities, or, which is the same thing, nearly as the height fallen by the ram to the head of the pile."

The following formula is employed by an eminent Engineer for the purpose of ascertaining the resistance of a driving or driven pile :—

Let $w.$ = Weight of ram } In Cwts.
$p.$ = Ditto of pile }
$h.$ = Height fallen through by ram, in inches.
$d.$ = Descent of pile by one blow of ram.

Then $\dfrac{w.\ 2h.}{2d.\ w. + p.}$ = The weight which will just sink the pile.

Having now minutely examined the mechanical construction and practical effects of these machines, we proceed with our inquiry relative to the degree of security afforded to the foundations of buildings by the application of piling. We esteem the opinion of those men whose genius, talents, and industry, aided by extensive practice, have contributed so powerfully to raise architecture to its distinguished position in the rank of sciences; and, upon the authority of one * (corroborated by the experience of ages) who stood proudly pre-eminent in mathematical skill, and its application to the above noble purposes, do we affirm, that the durability of piling, when properly employed, will outlast the durability of the superstructure itself. The premature decay that has occasionally occurred to the piling employed in many extensive works, has not arisen, in any instance, from the inefficiency of the material employed—that is, when the material (timber) has been selected from among the best of its kind. A *sine quæ non* with us in the judicious application of piling, is the most complete and perfect imbedding of the piles in the substratum; and also, the most effectual covering of their surface, by the masonry or other material being perfect in-bed, and well flushed with the mortar or other cement employed. Alternating exposures to heat and cold, dry and moist, together with the never-failing destructive effects of that

* Sir Christopher Wren.

most powerful of all menstrums, the atmosphere, are among the real and most active enemies of piling operations. Another and no less insiduous enemy exists in the "pile worm;" a creature that resolutely attacks the timbers, and speedily destroys them. A few years since the Author had an opportunity of examining the remains of some piling that had been destroyed by them: the piling was completely riddled with holes, and the interstices partially filled with sea-sand. The dikes of Holland have suffered severely from their ravages; but a method has been devised * for the protection of the piling, and appears to answer tolerably well. In this country Kyanised timber (as it is termed) has been extensively employed, and its beneficial effects, as a remedy for the dry-rot, have been most satisfactorily proved.

The most important applications of piling to the purposes of architectural construction may be classified under the following heads:—1st. Of piling for the foundations of bridge and other piers. 2nd. The application of piling to river, wharf, and embankment or retaining walls, including the varieties of close, grooved, plank, and inclined piling. The practice of piling for the support of such a cumbrous mass of materials as the bearing piers of a bridge, has been most generally observed, and is generally found to be adequate to the purpose. To confirm these observations, we shall instance at least three

* Philos. Trans., No. 455, Sect. 5.

important desiderata that are attained by a judicious application of piling:—1st. As a convenient and secure method of passing through a loose soil to one of sufficient density for the support of great masses of weighty materials 2nd. As a powerful assistant in consolidating loose or wet soil to a considerable depth. And, lastly, piling may be properly considered as the insertion of a firm strata opposed to an inferior one, by being placed at right angles to it. The piling and interstices between the pile, taken collectively, may be considered as a solid mass, whose friction against the sides of the surrounding earth will have a hold upon the soil, at least in proportion to its actual superfices. In a tenacious or compressed soil the amount of friction is the same for a single pile, fully demonstrating the security of this application of timber to the purposes of a most important branch of construction. The foundations of the bearing piers of Southwark iron bridge exhibit a very favourable illustration of the application of piling to pontile architecture. (See plate 2.) The piers were surrounded with a coffer-dam, constructed with three rows of piles of whole timber, enclosing a space elliptical on the plan. Each pier rests upon a massive timber platform, supported by piles; close to the outer edge of the offsets of the pier a row of timber sheeting piles were driven: this uniform belt of timber forms as it were a close stationary dam, preventing the substratum upon which the piles rest from being pressed outwards by the weight of the pier: a circumstance that frequently takes place where piling is employed, and the work heavy. This circumstance is well known to miners, and may be thus described: If a level *(in a mine)* be driven one or two hundred yards underground through the solid rock, there will be no danger of its not continuing entire for an indefinite length of time; but if the sides and roof only of the level be formed in the rock, and the bottom be cut through into a bed or substratum of clay, however strong and stubborn it may be, being pressed by the weight of the superincumbent rocks, it will by slow degrees swell and rise up in the level; and, unless it be continually pared down or prevented by some means, the level will, in no great length of time, be entirely choked up.* In tunnelling operations this tendency of the soft strata to swell and rise up is effectually counteracted by the inverted arch formed under the roadway of the tunnel. The Blisworth cutting on the Birmingham line of railway also exhibits an example of the inverted arch, adopted as a precautionary measure against the rising of the soil. In the latter work this mode of proceeding became the more necessary, as the rocky strata through which the cutting passes was underset or supported by retaining walls, in consequence of the intervention of the *weald clay*† which underlies the rocky strata. Close piling is executed

* Seaward on the rebuilding of London Bridge.

† This strata is subject to *slips*, carrying with it the superincumbent strata. The presence of moisture in this clay converts it into mud.

by driving two piles at a distance apart, these are connected together by bolting pieces of timber to them longitudinally, technically termed *wailing-pieces.* Grooved and plank piling is executed by grooving the piles, and introducing the edge of a plank of timber into the grooves. This latter description of piling is more frequently employed in the construction of coffer-dams, as it generally supersedes the necessity of puddling the joints with clay; it is not, however, remarkable for security. When the pier is completed (in bridge building) the piles forming the dam are withdrawn. The timber generally employed is either oak, elm, or beech. The coffer-dams of the new London Bridge were of an elliptical form, and consisted of three rows of piles dressed in the joints, without grooving: some of them measured between eighty and ninety feet: they were all puddled with clay, and shod with iron.

The foundations of the piers are raised upon piles of beech wood, on the heads of which were laid two rows of sleepers, 12 + 12 inches, covered with level planking, six inches in thickness. Another practice adopted in piling is, to lower a strong grating of timber, and between the openings or chequer-work of the framing the piles are inserted, and driven. Border piling is also driven round the exterior of this grating. An essential precaution is necessary in commencing the piling in this instance; namely, by driving the first piles in the centre of the framing, and proceeding with the work outwards, to prevent the earth becoming too compressed in the centre; which would be the case were the piling to commence at the outer row of interstices, or on the border of the grating.

In coffer-dams the larger piles face the water, and the double or triple rows are of smaller scantling. In some situations the grating of timber, covered with thick planks, and the pier carried up upon that, without piling, has been adopted. Plank or other piling driven obliquely into the earth is only bevelled on one side of the lower end; and such piles, if of small scantling, are charred on their extremities, instead of being shod with iron. The thickness of short piles may be allowed $= \frac{1}{12}$ of their length; but for long piles, the average diameter is 14 inches. Plank piles are in thickness from 3 to 5 inches, according to their lengths. It is a fact worthy of observation, that many of the failures occurring in buildings of magnitude, have arisen from the premature destruction and decay of their artificial foundations, and not from a subsidence of the soil! It therefore becomes a highly important consideration to trace the causes originating these destructive effects, and to suggest such remedies as appear to us based upon a philosophical and scientific view of the subject. The most pernicious effects have hitherto resulted from a practice which, by force of example alone, has gained considerable ascendency; we allude to the practice of raising the superstructure upon *planking*, or balks of timber laid flat upon the soil, without any further preparation than the mere levelling and ramming of the ground.

As the proximate cause of most failures arises from the subsidence of the substratum, the intervention of an artificial stratum of greater density is the most obvious means to remedy this defect. It is probable, therefore, that a stratum of sound timber laid horizontally was supposed to meet the difficulties of the case. Upon a slight consideration, it will be perceived that a more unscientific plan could not well have been adopted. In the employment of timber for the purpose of receiving or counteracting the effects of heavy strains, the force should be directed in such a manner as to cause compression of the fibres at the *extremities* of those fibres, or an extension of them by causing the force or thrust to be exerted in the direction of their length. A longitudinal framework of timber, laid horizontally and subjected to an enormous pressure by the weight of the superincumbent materials, will so far compress the fibres of the materials in the direction of their breadth, as to cause a serious and alarming disturbance in the equilibrium of the building; and eventually to ensure its destruction. The extensive settlements of the store-houses at Chatham and Woolwich, have, in a great measure, arisen from this reprehensible mode of construction. With a view to show the enormous weight that timber will bear when used as vertical columns or prisms, we here subjoin the crushing weight of different species of timber, deduced from an extensive set of experiments made by the late Mr. Rennie.

		lbs.	
Elm	Cubes of one inch	1284	Crushing weight.
White deal		1928	
English oak		3860	

Prisms of seasoned oak two inches square, and in height two, four, and six feet were crushed by vertical pressures of 17,500, 10,500, and 7,000 lbs. If four inches square, and of the same altitudes, the following loads were required to crush them; namely, 80,000, 70,000, and 50,000 lbs. Another vital objection to the practice of planking, &c., for foundations is, that a great portion of the timber remains exposed to the action of alternate damps, and the corresponding destructive influence of the atmosphere. Light sandy soils in low situations, and near large rivers, are particularly subject to the influence of the tides; filtrating upwards with the flux, and again becoming dry with the reflux: consequently the timbers are alternately subjected to dryness and moisture, by being placed on the surface of the ground, instead of being buried beneath it. On a clay soil the evil is not of less magnitude. The retentive nature of the clay will continually expose a wet surface to the *underside* of the timber, while the other surfaces will remain in various states and temperatures: decay rapidly ensues, and the building fails, by the premature destruction of the very means adopted for its preservation. While we justly condemn this mode of building, on defect of principle, we are bound to mention that there is every reason to hope that the preservation of timber from the premature effects of the rot has been effectually accomplished

To counteract the destructive influence of the atmosphere, and to secure timber from the effects resulting from exposed situations, has occupied the attention of the scientific experimentalist for a long succession of years. The antiseptic qualities of coal-tar has been long known, and in many instances, where perseveringly applied, has been productive of considerable benefit. Although the recent invention of Mr. Kyan would appear to leave nothing to be desired, yet there are still many who place considerable reliance on the former method of preserving wood and metal. The simplicity of the operation, and the cheapness of the materials employed, are urged as additional recommendations to the use of the anti-corrosive composition. In the year 1822 Mr. John Oxford procured letters patent for an improved method of preventing "Decay in Timber, Metallic Substances, and Canvass, &c." This was proposed to be accomplished by the following method:—The essential oil of the tar is first extracted by distillation, and at the same time saturated with chlorine gas. Certain proportions * of oxide of lead, carbonate of lime, and carbon of

* See specification in the Repository of Arts, Vol. lxiii., N. S.

purified coal-tar, are well ground, and mixed with the oil; the composition is then applied in thick coatings to the substances intended to be preserved. The penetrative qualities of this compound are considerable, and may be employed with advantage in many cases. An American patent has very recently been granted for a similar purpose; and the ingredients of the antiseptic fluid are nearly identical with those already described, the metallic oxide, however, consists of iron. Since the year 1812, a gentleman of the name of Kyan prosecuted an immense number of experiments on the preservation of timber; and "at length ascertained that *albumen* was the primary cause of putrefactive fermentation, and, subsequently, of the decomposition of vegetable matter." The affinity of corrosive sublimate for this material was not unknown; but the credit is due to Mr. Kyan for successfully following up by a well conducted series of experiments, and ultimately deciding in a most satisfactory manner, that vegetable matter may, by saturation with the bichloride of mercury, be effectually made to resist the effects of putrescent fermentation. This therefore forms the basis of an invention * which is likely to prove, above all others of a similar kind, a most powerful agent for the indefinite preservation of a great variety of vegetable substances. Its application to the preservation of

* Now in the hands of a respectable company, called "The Anti Dry Rot Company."

timber may justly be ranked among the most important benefits conferred upon society by the hand of science. The application of iron for the purpose of piling has recently been adopted in some situations that are likely to prove severe tests of its applicability, and, if successful, to warrant a more extended use of this material in a variety of situations where timber piling might be considered objectionable from its exposed position. In an admirable memoir on the subject of iron piling, published in the first volume of the "Transactions of the Institution of Civil Engineers," it is therein stated, that the introduction of metal for the purpose of piling originated with Mr. Mathews, of Bridlington, previous to the year 1822. A gentleman of the name of Ewart (now of Her Majesty's Dock Yard at Woolwich) secured by patent, in 1822, a plan for the construction of iron piling, applicable to the formation of coffer-dams. It appears that this invention was not originally contemplated for the purposes of permanent piling; notwithstanding which, subsequent adaptations to the latter purpose lead us to infer, that they are all derivable from the mechanical arrangement exhibited in the plans of Messrs. Mathews and Ewart. The priority of invention, however, is most justly due to the former gentleman; Mr. Ewart having matured his invention without a previous knowledge of the proceedings of Mr. Mathews.

Mr. Ewart's plan has been most successfully

carried into execution, under the able direction of Mr. Milne, the Engineer of the New River Company, at Broken Wharf, and by Mr. Hartly, of the Liverpool Docks, at the basin of George's Dock at Liverpool. A detailed illustration of the construction of the piling is shown in plate 3,* which is fully explained in the accompanying reference to the various figures. Fig. 1, is a plan of three metal piles, united together by cramps of cast-iron. The cramps B are made to fit and slide easily on the dovetailed ribs or ledges *a*, when the piles are placed side by side, yet sufficiently dovetailed to retain the piles or plates in their proper situations in relation to the cramps. Fig. 2 is a plan of the heads or upper parts of the piles, and cross sections of the cramps for securing them at their vertical joinings. Fig. 3 is a back elevation of the piles as seen in the *inside* of the coffer-dam, showing also the horizontal joinings where additional lengths of piles are required. Fig. 4 is an end view of one portion of a pile, detached at the horizontal joint. Figs. 5 and 6 are views of the cramp and wedge for securing the piles at their joints. Fig. 7 is a side view of one of the vertical cramps B. Fig. 8 is a view of that part of the same cramp which is applied next to the piles. Fig. 9 is a cross section of the lower part of the same cramp, through the line *o p*. Fig. 10 is a horizontal plan, of the form of a coffer-dam, for the

* Collated principally from the Specifications given in the 43rd volume of the Repertory of Patent Inventions.

foundation of a pier. Fig. 11 is a plan of part of a coffer-dam, showing how the piles are arranged round it. Fig. 12 is an end view or elevation of part of the same. Fig. 13 is a plan of three piles, the middle one being curved to admit of expansion and contraction in the line of piles on each side of it. Fig. 14 is a horizontal section of two cast-metal piles, united at their vertical joints by cramps cast in one piece with the piles. Fig. 15 is a horizontal section of three piles of wrought-iron, united at their vertical joints by cramps of wrought-iron or cast-metal. The patentee states, in his specification, that the piles are made in portions of various lengths, generally from 10 to 15 feet, but shorter or longer when required. The vertical cramps are also made of various lengths, corresponding to the lengths of the piles, along which they are continued from top to bottom; but so that the joints of the cramps, end to end, shall never be in the same horizontal line with the joints of the piles which they connect. When necessary, these cramps are strengthened by making them thicker than shown in the drawings, or by adding a rib to them as represented at *c*, Fig. 1. The construction of this kind of piling has been successfully modified in its application to various works, by Messrs. Walker, Cubitt, Sibley, and others; the plan devised by Mr. Sibley exhibits the greatest deviation from the original idea; and, in fact, may be almost considered as a distinct invention. These piles were made hollow, and when driven, were filled with *concrete.* With respect to the effect produced by the pile-engine in driving iron piling, it may be justly inferred that considerable care and attention must be requisite, in order that the blow of the ram should fall with its due effective force upon the exact centre of the pile-head. The great danger of striking piles of timber obliquely, when it is the intention of the engineer to drive them at right angles to the horizon, has been strikingly exemplified at Westminster-Bridge, where the rupture of two arches of that structure was occasioned by a defect of this kind. The iron piles at the Brunswick Wharf were driven with a ram weighing from 13 to 15 cwt., with a fall of three feet six inches only; in some cases the fall was even limited to a less extent. The crab-engine was invariably used for this purpose. The tendency of iron piles to start from their position when struck with a heavy blow, suggested the propriety of driving them by a series of smart shocks given in succession. By these means the driving may be conducted with great security and facility, care being taken in the first instance to cap the head of the pile during the operation with a sound piece or plank of timber, to transmit the shock more equally on the metallic surface.*

* For a more elaborate detail of the preparations required for driving iron piling, such as boring the ground to receive the pile, cutting trenches, &c., &c., the reader is referred to the first Volume of the "Transactions of the Institution of Civil Engineers," published by Weale, of Holborn, 1836.

We shall now advert to the comparative permanency of iron piling. It has been practically demonstrated, that iron, buried in the earth under water, and kept perfectly free from atmospheric influence, will not sensibly corrode for an indefinite length of time: indeed it may be safely affirmed that nothing short of volcanic agency or great intestinal commotions and changes, which in most cases occur only at intervals of great distance of time, will produce a decomposing effect upon iron thus situated. A change, however, occurs in the material; for it has been proved that iron thus buried, will, in process of time, be converted into *steel* of excellent quality. The action of *water* upon iron effects a result totally different, a certain chemical combination taking place, producing, within a comparatively short period, a material resembling plumbago. This effect also takes place upon the pistons of steam engines that have been subjected to much work, the interior parts, consisting of the wedges, &c., being readily cut with a knife. These remarkable facts being known, it becomes an enquiry of great practical importance to ascertain the cause or causes of such change of structure in this metal, and also to discover some new compound that will effectually resist the decomposing action of water upon it. Experiments have not hitherto been made to that extent as would offer sufficient data for a more close chemical investigation of this subject. Until this be effected, the practice of iron piling must be regarded with a much less degree of reliance than that of timber piling, secured by saturation with the bichloride of mercury. In addition to the perpendicular pressure acting upon the horizontal surfaces of piling, an enormous lateral thrust is to be resisted in cases where river, sea-walls, or embankments are raised upon piling. Considerable care and attention is likewise requisite, in order to give to the masonry that degree of facial curvature most capable of resisting the effects of a lateral thrust. The subjoined method has been employed with uniform success in works where the thrust has been great, or where land-slips have been apprehended. The following is the construction of the diagram shown in plate 4. Let A B, B C, be given, join A C, and bisect it in D, make D G = C D ÷ 4, join C G, G A, bisect them at I and K, let H I, K J = $\frac{DG}{4}$ join C J, J G, G H, H A, and so on *ad infinitum*.

C, J, G, H, A, are points in a curve, that is a tangent to A B, and projects B C.

Fig. 2 represents a retaining wall, with a curvature produced from the foregoing formula. River, wharf, and sea-walls have occasionally to withstand violent forces acting from without, caused by mooring vessels thereto, or, by warping them in and out of harbours, docks, and basins. The mooring-rings are secured to the iron land-ties, and these latter are firmly bolted to piles driven into the ground at the back of the wall, in a slightly inclined position. The enormous horizontal strain to which these piles are subjected, frequently cause them to yield; and consequently the whole force is transferred to the backing of the wall, which is mainly protected from being pulled outwards by the security afforded by its batter, or slope, and the strength of the fender piling, with which the outer work is generally faced. As an improvement on this method, we would suggest the application of Mr. Mitchell's "Patent Screw Mooring," fixed horizontally, and to which the tension rod, forming the land-tie, might be firmly fixed. A screw mooring of 3 feet diameter would thus have a secure hold of an area not less than ten superficial feet.* There are certain circumstances connected with the practical operations of pile driving that require a passing notice.

1st. It has been frequently found that when piles have been driven to the depth that was at first deemed desirable, another blow with the ram will, to the great surprise of the engineer, suddenly sink the pile to a greater depth than was effected in the first instance This effect may arise from several causes: among the most probable of these, may be mentioned the sudden transition from a dense to a looser soil, or the intervention of a stone immediately under the foot of the pile, where the immediate strata is known to be sand. Now a smart blow, accidentally given in an *oblique* direction, will cause

* For drawings, &c. of this excellent invention, see "The Public Works of Great Britain," and "Mechanic's Magazine," No. 756, Vol. 28

the pile to swerve from its original upright position, and to pass still further into the strata, that was at first supposed to be sufficiently consolidated to resist the further efforts of the pile machine. Again: when the head of the pile has been much bruised by the action of the ram, the force becomes deadened, and ultimately nullified by the comparative softness of the abraded surface. In this case the top of the pile should be cut off, and the driving continued on the new and firm surface thus exposed to the ram. Mr. Timperly* argues from these facts, that all theories must be uncertain which profess "to ascertain the actual weight a pile will bear, by having given the weight of the ram, the fall, and the depth driven at a stroke." This, however, is not the case, as the theory has for one of its principal conditions, the supposed uniformity of the strata, through which the pile is intended to be driven.

The most conclusive evidence in favour of the enormous friction on the sides of a pile, and the consequent degree of security to be relied thereon, is shown by the immense power required to withdraw them from their position, when once firmly driven into the soil. The gentleman above alluded to has given a particular statement of some operations of this kind. "In removing the temporary bridges and coffer-dams,† the piles were principally drawn by

* Account of the Hull Docks. 1st vol. Trans. Inst. C. E.

† Hull Docks. *Ibid.*

the engine crabs,* with double blocks and chains, and so firmly did they hold, that some of them required sixteen men with four crabs to remove them, but in general half this power was sufficient; after the piles were started, one crab with four men (assisted by the buoyancy of the water†) accomplished the business. The power applied to some of these piles was not less than from fifteen to twenty tons. There being occasion in the course of the week to draw several of the sheeting piles in the Whitefriar-gate lock pit, a 4-inch screw was used, and one of the piles, 14 feet long by 12 inches wide, required, on the most moderate calculation, a power of 18 tons to draw it, the soil being nearly a pure sand; another pile could not be drawn by even a greater force, until a hole was dug around it, but the others, being in softer ground, moved more easily." It is further stated, that some of the permanent bearing piles sustain a weight = 20 tons, without settlement.

The operation of piling requires so much heavy timber, and is withal of so expensive a nature, that effective machinery for withdrawing piles from the bed of rivers is a great desideratum. The timber may afterwards be employed in farther service, or in works of the same description. In 1747 Marshal Belleisle passed the Var to the army of the Allies,

* The construction of a crab-engine is described further on.

† Query? Weight of the water.

for the purpose of conquering an adjacent province (Nice); and to facilitate the communication with France, he caused to be constructed across that river two magnificent bridges, measuring in length about 300 toises each, and of sufficient solidity to resist the force of one of the strongest currents in Europe, at a time, too, when the river was most subject to them. When the evacuation took place in 1749, the bridges were destroyed. The foundations, to use the words of M. Belidor, comprised a "forest of piles." This writer communicated with M. Guil, Brigadier-General of the king's army, the design of the machines shewn in plate 5, figs. 1 and 2, as it was deemed necessary to withdraw the whole of the piles, many of which were buried at the depth of 12, 13, and 15 feet in the bed of the river, and in an extremely tenacious earth. After repeated trials to no effect, arising from the cords breaking, they employed nine to each machine, and the piles were then drawn with great expedition, each in less that four or five minutes.*

The general construction of this machine is explained by figs. 1 and 2. A beam, G L, is placed upon a support or fulcrum, B; the end of the beam G is bound with iron straps, having a hook, Z, to lay hold of the cord or chain, which is fastened to the pile C, by means of a short bar or rod of iron driven through the pile; the beam is elevated for this purpose by the windlass P, and when made fast

* Belidor Architecture Hydraulique, Vol. I. p. 121, 122.

at one end (Z) is released at L: consequently the beam is allowed to fall by its own gravity, drawing the pile C upwards. The length of the beam, in this case, is supposed to be 18 feet, and the length, G K, two feet; but in all cases, whatever the length of the beam may be, divide the same into nine equal parts, and allow one part for the length from the extremity to the fulcrum at K. The action of the apparatus may be rendered more powerful by the workmen mounting upon the end, L, and thrusting it downwards.

The substance of M. Belidor's calculation of the power of this machine is as follows:—Let us suppose the beam to be 18 feet long and 12 + 12 inches, and, as we have before stated, G K will be 2 feet long, and K L 16 feet in length. Now a cubic foot of dry oak weighs about 60 lbs.; let us take this as the basis of the formula. For greater accuracy, make K H = G K, in order to consider these two parts of the beam as being equilibrated. Imagine that the rest of the weight H L is united in the middle A, the centre of gravity of the part H L to take the place of the power which is to act upon the pile; recollecting that this power will be = 840 lbs., which is the weight of 14 cubic feet of dry oak, answering to the arm of the lever K A, 9 feet, whilst the K G, in regard to its effect, is only 2 feet. Thus, in the state of equilibrium, the power A of 840 lbs. will be, to its action in drawing the pile, as K G is to K A, or as two to nine.

Following the rule, we shall find that this action = 3780 lbs., which can be augmented by the thrusts given to it by the workmen mounted upon the end, R L. If we suppose four men acting together by supporting themselves by the cords D D, fig. 2, suspended from the top of the machine to prevent them falling, it will appear, that weighing together 600 lbs. at the extremity of the lever K Q, which is equal to six times K G, their effect upon the pile will be 3600 lbs., which, added to 3780, gives 7380 lbs for the force with which the pile C will be drawn upwards. The power of the machine may again be indefinitely increased, should the pile refuse to move from these exertions: displace the pullies from their position, M N, and fix a ring, F, to the end of the beam, which will be kept raised by the hook Z laying hold of the chain or rope attached to the pile; a pully is then fixed at the foot of the machine, and a cord or rope passed through the ring F and the pully at the bottom; the workmen will then pull horizontally at the bottom, and draw the beam downwards with great power.

This machine may be also made to answer the purpose of the ordinary pile machine, by fixing a heavy ram at one end; it is then raised and allowed to fall as usual, or might be worked after the manner of a tilt-hammer. Fig. 3 is another apparatus for withdrawing piling; C is a screw supported in a square timber framing, worked with the levers B A in the manner of a capstan. The pile D is attached to the hook at the bottom by cords or chains, and is of course drawn upwards. This may be readily removed by boats from one place to another. Its principal defect is, that if the pile is very firmly fixed, the stand will not remain so steady as might be desirable for the purpose of its operation. The weighing machine for withdrawing piles simply, consists of a flat-bottomed craft or barge, having at one end a windlass, and at the other a strong chain, which is also fixed round the pile intended to be raised. In the centre of the barge, and laid longitudinally, are two stout planks or beams, forming a tram road for a truck, loaded to the amount of several tons weight; the truck is made fast to the chain coiled round the drum of the windlass. The mode of operation is as follows:—the truck is first drawn to that end of the barge nearest the pile, which considerably depresses the head of the barge in the water; the chain is then fixed round the pile; the windlass is now, in its turn, put into operation, drawing the truck towards the opposite extremity of the barge; the consequence is, that the end becomes depressed, and raises the pile at the other end.

At Waterloo Bridge, Bramah's pump was used for the purpose of withdrawing the piling. Fig. 4, plate 5 shews the arrangement employed: A A exhibits part of the piling of the coffer-dam; B is the pump fixed upon the top of the piling; by working the lever, or handle C, water is withdrawn from the cistern D, and forced through the copper pipe E, into an iron cylinder enclosed within a casing of wood, F, firmly bolted to the piles; the solid plunger G is thus forced upwards, and raises a beam, H, placed upon the fulcrum I; a loose ring of iron, L, is dropped over the pile K, and the chain M is passed several times round the beam H. The loose ring L being drawn upwards, will shift itself into an angular position, and will take a purchase upon the sides of the piles, causing it to follow with the elevation of the beam H, until it is completely withdrawn from its position.

We shall now proceed with an examination of those artificial substrata, which, by their peculiar preparation and properties, are included in the class or genera of concreting mixtures. The first that claims our most particular attention is the *bèton* of the French. This substance has been (erroneously we think) considered by several writers as identical with our *Concrete* and the *Smalto* of the Italians. In order that the reader may fully comprehend the nature and properties of this important artificial conglomerate, we offer in the following chapter a translation from the highly valuable work of M. Belidor,* containing a detailed analysis of its composition and use.

The composition called *bèton* is understood (in a popular sense) to be a kind of mortar used by the French Engineers, consisting of the following

* Architecture Hydraulique, vol. 4, section 2. De la manière d'employer le bèton pour les fondations faites au fond de l'eau sans batardeaux ni épuisemens.

materials and their proportions. Twelve parts of *pozzolano*, or Dutch *terrass*, six of good sand, nine of unslacked lime, the best that can be had, thirteen of stone scraplings, not exceeding the size of an egg, and three parts of iron scales from the forge. These, being well worked together, must be left standing for twenty-four hours, or till it becomes so hard as not to be separated without a pickaxe. The mortar being thus prepared, a bed of rubble stone, not very large, is thrown into the coffer, and spread all over the bottom as nearly level as possible. They then sink a box full of this hard mortar broken into pieces, till it comes within a little of the bottom; the box is so contrived as to be overset or turned upside down at any depth; which being done, the pieces of mortar soften, and so fill up the vacant space between the stones; by these means as much of it is deposited as will form a bed of about twelve inches deep; another bed of stone rubble is now thrown in, and thus one layer of *bèton* and one of stone is alternately deposited till the work approaches near the surface of the water, when it is levelled to receive the regular courses of masonry. The text minutely describes the precautions necessary in the erection of large works, and the general method of the proceedings adopted by the most scientific engineers.

CHAPTER III.

ON THE NATURE AND APPLICATION OF BETON.

FROM THE FRENCH OF M. BELIDOR.

To initiate those who have had but little experience in works such as these in question, we shall preface our observations with a general and popular view of the subject, that they may the more clearly comprehend the detailed description that we intend offering to their consideration. With this view, therefore, let us suppose that the services of the architect or engineer are required to direct the construction of the foundations for a fort or jetty in the sea, at the depth of fifteen, twenty, or even twenty-five feet. The works should be commenced by forming the necessary boundaries to mark the extent of the proposed erections. The usual examinations of the nature of the ground, the state of the levels, &c., need not in this particular instance be made, for the space so enclosed may be permanently raised, and completed at leisure, by filling it in with such materials as are hereafter described. The foundations, therefore, need not extend beyond the enclosure required for the walls, care being taken that they have a thickness proportionable to the depth of the water. The interior surface being perfectly level, and the exterior having a slope of a fifth or sixth of its height, as in common *revètemens,* in order to avoid unnecessary toil and expense. The works being marked out by buoys, and guarded from the action of the waves by a temporary counter mole, if judged necessary, are commenced, by erecting a coffer-dam of the usual construction, by piling, &c.; this dam, however, need not be so complete in its appliances as those generally employed for the purposes of bridge building, where the water is required to be withdrawn from the interior of the dam, but is nevertheless to be constructed with such regard to security as will enable it to withstand the force, &c. of water from without, and the pressure of materials, water, &c. from within. The water of the interior will consequently be displaced by throwing in the *bèton,* with such apparatus as we shall presently describe. It being of the first importance, that the cement should subside upon the natural and consolidated layer, or strata, means must be employed to clear the bed of the river or sea from all extraneous deposits: this can be effected by removing the ground to the necessary depth, with a *machine à cuilliere* placed on the summit of the coffer, which may be advanced as the work proceeds, and will lay open the ground to the depth of the firm strata, which, in most instances, will be but a few feet below the alluvial deposit. Should the strata itself be found to consist of a muddy, loose, or friable quality, the common precautions must be resorted to.

Successive masses of *bèton* may now be thrown into the coffer, which may be filled to low water mark; but as this mortar, in falling, would nevertheless have to pass through a great head of water, it would lose much of its value by the greater part of the lime being washed away, means have therefore been found, by a certain contrivance hereafter mentioned, to lower it without the risk of having its properties destroyed. Although the coffer intended to contain the masonry in question be very simple in its construction, our description will be rendered more intelligible, by reference to the subjoined illustrations, which represent the coffer in its working state.

Fig. 1, plate 6, represents a side elevation; the letters A B show the high water mark, and C D the bed of the sea. The piles A B belong to the ranges driven on both sides, to form the *encaissement* for a distance of no more than 10 or 12 toises (so as not to encumber the work which, in most instances, require to be executed with extraordinary expedition). They are headed with a cap D, to which the higher pieces E are fixed, and the lower, G, suspended by holdfasts or irons attached to H, to support the timbers F, which compose the coffer, on whose edge two thick planks, O, serving as a platform to the rollers, R, carrying the machine for throwing in the *bèton*. The line I K marks the bottom of the trench cut below the bed of the sea C D; and, lastly, the sides I, L, M, R, enclose the *bèton* mixed with stones, wherewith the foundations are commenced. Fig. 1, plate 7, represents a plan of the coffer and machine, another elevation, which will be found in Fig. 1, plate 8 : thus these illustrations, studied together, with reference to the letters which mark the same throughout, leave no part of the subject unexplained.

The width of the coffer being equal to the thickness of the masonry, the rule is, when carrying on works such as operations in the interior of a haven, to calculate this thickness at one-half of the depth of the water, but should the masonry be exposed to the open sea, or a strong current, then it ought to be two-thirds. By this rule, also, you will judge of the proper distance between the two ranges of piles, the strength of which will be proportionate to the height of masonry and *bèton*, and to the depth they must be driven so as to penetrate some feet in firm ground; thus rendering them capable of sustaining the impetus of the waves, and preventing the masonry from spreading until it has acquired solidity.

On this account, according to the nature of the soil, they must be armed with iron *(lardoirs et frettes)*, and the timbers F, cut off at three or four feet above the level of the highest tides, so as to secure the work against the action of the waves, when it shall be sufficiently raised to encounter them. These piles should be distanced according to the height of the masonry, so that their common resistance may be proportionate to the power they will have to overcome from its spreading; always regulated, however, according to its height; whenever that does not exceed 10 feet, 6 feet from the centre of one pile to the centre of the next will be sufficient. At about 16 feet they should be 4 feet apart, but when it rises to the height of 20 or 25 feet, the distance of 3 feet apart should not be exceeded. All the piles should be securely bound together by a cap, the two opposite rows being keyed together at intervals; and, having cross quarters traversing the top of the coffer, the more effectually to secure it against opening; these may be removed when the *bèton* is consolidated, and completed by its uppermost course of common masonry. These piles will be again bound by two bands* placed internally, so as to serve for supports to the *vanages*, the upper one a great deal thicker than the lower, so as to admit sufficient space for the sloping of the plank piling, as much on one side as the other (which is seen in the front elevation), wherever a pier or isolated jetty is to be constructed.

But if the question refer to the casing of a quay, or the foundation of a breakwater, then the slope need only extend along the exterior side, the interior being perpendicular, so as to allow for the pressure that the masonry will have to sustain from the nature of the soil. Thus it is readily seen, that

* Wailing pieces.

the two corresponding heads on the inner side should be equal, or nearly so, according to the intended slope, care being taken to place the second a third or fourth of the water's depth, according to the height of the coffer; the only means of fixing these are by suspension, with iron holdfasts, or ropes attached to piles above high water, as may be seen at H, thereby affording space to work. Respecting the plank-piling there is a visible necessity for making them, not only of a thickness proportionate to the height of water, and spread of masonry, but also to the depth which may be considered necessary for the trench (cut to carry off the sand at the bottom of the *encaissement* from the firm ground), into which it is necessary that they should be driven at least 2 feet. These observations are equally applicable with respect to the piles, the quality of the soil should be examined by frequent borings before the work is commenced, which will, at the same time, determine the amount of labour required for this portion of the operations.

Further, we shall merely premise that the plank-piling should be well bound together by grooves and wedges, according to the usual practice, and cut down level with the cap of the piles, each being fastened to the bands, leaving no aperture, however small, at the joints, in order to secure every particle of the lime and *pozzolano* within the proper limits, so as to preserve the masonry free from flaws: this being accomplished, the *machine à cuillière* will be required to clear away the sand.

We may observe *en passant* that the foregoing description of a good *encaissement* is applicable to coffer-dams, erected when it is requisite to drain some part of the sea, to enable us to clear away the remains of ancient sunken works, which operation sometimes costs more than that of their original erection; this proves the necessity of making marine-works perfectly solid, taking into consideration every possible circumstance that might happen to cause premature decay, or the chance of a remote failure. Our knowledge of antique monuments evidently proves, that with good materials, it is always the fault of those who know not how to employ them properly, that their works are far from promising a glorious destiny, falling for the greater part to ruin during the life-time of their constructor. The pieces forming the base of the machine in question, consist only of three soles or logs A B, fastened to the transverse beams C D, the two first being again strengthened by the cross quarters E F. In the engraving these are only dotted in like the rest, the base being partly covered by the upper parts, but it is easy to imagine it, particularly as it is shown in the side elevation, where the frame G H I K is repre-

* Dredging Machine ?—(*Translator*)

sented raised on the third sole A B; this frame serving, with two others like it, to sustain the axle of the wheel and *cuillière;* thus a little examination is sufficient to explain it.

We now proceed to the action of the machine.

A slight description will prove sufficient, as it is constructed on the model of those employed for clearing Toulon harbour; we shall merely premise that all the apparatus is carried on two small beams, *a*, *b*, placed across the top of the *encaissement*, and progressing gradually, after having cut one part of the trench as far as required. The handle of its *cuillière* passes between two rollers L M, against which it works, viz. against the first, L, in descending, and the second, M, when ascending. This motion is caused by the action of the wheel, and the two principal ropes attached to the *cuillière;* the first N Z Y causes it to dig or penetrate in the ooze, and raises it again when full; this is done by turning the wheel from right to left. The second, T V S O, (the *tire arriere*) passing over the pulley S, and against the rollers V, causes the *cuillière* or scoop to descend, after having emptied its contents into a mud-boat, by opening the trap or valve, supposed to be situated at Q, which is done by turning the wheel in a contrary direction. Towards the extremity of the handle, a third rope O P is bound, the other end of which is belayed three or four times to the third sole G K to steady the *cuillière* by means of the fastening at P; to drive it into the ground, according to the progress of the wheel, and to direct it so as to dig equally over the whole width of the bottom; when filled it can be raised by slackening the rope. Lastly there is a fourth cord X, passing over the pulley x, with which the foregoing operation is sometimes effected, so as to assist in raising the *cuillière*, when laden more than usual.

It is necessary to observe that this machine can not only be used for raising the ooze from the bed of the sea, but also on all other occasions where it is necessary to excavate deep trenches in marshy lands, moss, or moving bogs, in order to erect embankments or causeways cased with masonry, &c. &c.

Two parallel rows of plank piling must first be driven into the earth, the length and thickness of them being proportioned to the depth that they are required to be driven. They should be driven as far as possible into the firm ground, and the loose soil should then be removed entirely from the *encaissement* or enclosure. One or more dredging machines must now be employed to excavate the remaining portion of the trench, until a secure and firm substratum is obtained.

The entrance of the water into the enclosure is of no immediate consequence, if sufficient care is taken to sustain the exterior pressure by a judicious arrangement of the cross quarters, keys, framing, &c. Having obtained a good substratum, it must be cleared as much as possible, and a gentle slope given to that side opposite to the facing of stone, to secure the foundations against the pressure that the lining will have to sustain, taking care that it be sufficiently wide to dispense with counterforts, as it is probable that considerable difficulty might arise in attempting to excavate short transverse trenches, from the little play that the scoop could possibly have. It is not requisite to level the ground with any great degree of accuracy; even the masonry employed does not require it; on the contrary, the little interstices that remain contributes to bind the work more firmly together. If the firm ground is discovered to lie very low, so that you are compelled to build upon ground of an inferior degree of consistency, then it is only necessary to dig six or seven feet below the bottom of the water; after which drive piles, placed chequerwise, three feet apart, as far as you can with a heavy beetle; let their heads reach just four feet above the ground at the bottom of the foundations, without levelling them, which can even be done in the water by the aid of *false stakes;* then cover these piles, and it will form a compact mass. But if the ground should prove marshy, you must excavate as deep as possible, sink a timber grating, and drive the piles in the reticulations leaving them jutting as before; throw in the *bèton* as usual, and level it carefully over the entire surface, so as not to load the grating more on one side than on the other.

It is of the utmost importance that the plank piling employed for the coffer-dam should be of the maximum degree of strength, well bound together by redoubled wailing. The piles should be driven as far as possible into the firm ground, to prevent so soft a soil, as we have supposed, from giving away when laden with the superincumbent masonry. Observe, that the weight of the masonry should contribute as much as possible to its firm consolidation; the greater portion of the water with which the *bèton* is saturated, being expressed or forced out by this means. It must be acknowledged, that by placing the foundation thus, there can be no obstacles presented by the state of the ground, but what may be successfully overcome; in fact it is doubtful that the before-mentioned cases could be otherwise accomplished, for there can be no question about the necessity of the dams, the erection of which might be as difficult as that of the principal work. Would it not be easy to prevent the leakage, and keep the foundations continually dry? To finish the description of the drawings, explanative of the machinery for laying or immersing the *bèton*, it only remains to be observed, that it consists of two uprights raised on a frame Y, carried on two rollers R, serving to move it along the *encaissement:* these uprights, which are bound together by two tie-beams S T, substain the axle, from whence is suspended the chest W, by two ropes always kept equidistant by two rollers X, moving on one of the tie-beams. This chest is three feet square; its bottom Z *b* is made to move in the manner of a pump valve, on two pivots *b*, and fastened when shut by a catch *m*, below which to a ring Z is attached a cord *a* Z, made fast to an iron pin *a*; the length of this cord must be regulated, so that the case being let down to within about four feet of the bottom, can go no lower without its weight forcing the catch to let go the bottom, which then opening, permits the *bèton* to fall out.

This bottom is supported on two sides, by the ends of the chains *n* fastened to the lower edge of the chest, to support it in the position as it appears (dotted) in the plate. When by reversing the motion of the rollers the chest rises, then re-fasten the catch by drawing the attached cord *r l d*; this is done when out of the water by pressing the handle *b d*, the catch *m* may then be loosed, thereby enabling it to resume its vertical position, and to sustain the bottom as before. We may now raise the chest to a convenient height, and re-fill it: to effect this, the wheel must be fastened by tying the cord *f* to part of the frame. It is requisite that the inside of the chest should be very smooth, and the joints well caulked; that it be pitched without, and shut with a ground lid; that the column of water displaced by its descent may not carry away any of the *bèton*. A layer of gravel is required to be laid in the chest before filling it, in order that the *bèton* or cement may not adhere to the bottom. The interior of this chest, which is three feet cube, contains twenty-seven cubic feet of *bèton*. It may be made larger, and in proportion to the foundation you have to make;

nevertheless, (for fear of retarding the action of the apparatus) it should not exceed four feet cube, which will give a solid content of sixty-four cubic feet.

Vitruvius, speaking of *bèton,* says, "it should be composed of two parts of *pozzolano* to one of lime." He undoubtedly means the latter to be measured unslacked, and slacked when the cement is made; so that the consequent effervescence may throw off a greater quantity of the salts contained in the *brocaille,* and the incorporated *pozzolano.* It is further supposed, that this substance is equally attainable as in Italy. M. Milet de Montville, who has acquired (in the kind of building we are treating of) an experience enlightened by good principles, has made numerous experiments on the best manner of compounding the *bèton*—having proved it most successfully in the works which he has superintended. Having chosen an even and well-beaten ground, take twelve parts of *pozzolano, (Terrasse de Hollande,* or *Cendrè de Tourney)* of which you will form a circular wall of five or six feet in diameter, on which place six parts of sand well sifted, free from earthy matter, and evenly spread. Fill the interior of this circle with nine parts of quick lime, well calcined and pulverized with an iron beetle, and to cause it to slack more quickly (in maritime works) throw on sea-water in small quantities, stirring it from time to time with an iron spatula. As soon as it is reduced to a paste, incorporate the *pozzolano* and the sand.

The whole being well mixed, throw in thirteen parts of unhewn stone, and three parts of iron dross well pounded. If this latter ingredient cannot be obtained, sixteen parts (instead of thirteen parts) of rough stones or pebbles must be added, of a size not larger than a pullet's egg. Let this composition be well amalgamated for the space of an hour, after which it must be left in heaps to coagulate; for this purpose the space of twenty-four hours will be sufficient in summer or in warm climates, but in winter it often requires the space of three or four days. Observe to keep it protected from the rain, and not to use it until it has sufficiently hardened to require breaking with a pickaxe. It must be borne in mind, that, though the *bèton* is employed in a consolidated state, when it has passed through the water and has reached the bottom it spreads and settles. The plunging machine is advanced, according as a bed of ten or twelve inches in thickness is spread over the whole foundation; after which place unhewn stones of moderate size, the largest not exceeding the fourth part of a cubic foot. These stones should be carefully arranged side by side: they will sink into the mortar which is now in a soft state, but which will, in the space of three or four months afterwards, become indissolubly united, and its firmness will increase with its age. This bed of stones having been again covered with a fresh stratum of *bèton*, commence another, and so on alternately, even to within six or seven feet of the surface of the water.

The use of the chest may now be dispensed with, and the *bèton* may be thrown in with buckets or baskets, care being taken that they be emptied very near the surface of the water,—it being essential to prevent the dilution of the cement, which would otherwise be the case if it had to pass through a great head of water; for, the heaviest parts sinking more rapidly than the others, all the strength of the lime would be lost. It is evident that, by this method, the masonry may be looked upon as being done according to the best rules of the art, as it is composed of very small portions of materials, presenting at the same time a large surface to the *bèton*—therefore, a great quantity of the salts necessary to the concretion are detached. It has also this convenience, that it may be laid without the assistance of a mason, the simplest means only being required to place it in its proper position,—all the workmen required for this purpose being a few carpenters to make the chest: besides, it is better to keep the chest in which the artificial masonry has been made, particularly in places exposed to the waves of the sea or a rapid current of water. By the foregoing method one of the jetties at Toulon was constructed in the year 1748.

The engineers of this place not being acquainted with any method more available for fixing solid masonry in the sea without vexatious impediments, it may be justly considered the best method of ensuring the most complete success. It is not to be apprehended that a facing stone, being detached, would be followed by many others, and that all in the water would fall successively to ruin; and let us ask, should we not calculate greatly on the economy of being able to work without dams or draining, which sometimes cost as much as the work itself? Is there not reason to be surprised, that a practice of which the ancients made such good use should be so little employed, except on the Mediterranean coasts? Nevertheless, it is equally applicable to seas and rivers, to erect a quay, or the piers of a bridge, in places which are never dry, and where there is always a great head of water; being preferable, in most cases, to foundations made in "*caissoons*" laid along the bottom of the river.

M. Milet de Montville having filled a chest containing 27 cubic feet of *bèton*, sunk it in the sea, where it remained during two months, after which it was drawn up to ascertain the consolidation it had acquired. On inspection it was found to be converted into so compact a body that more difficulty was experienced in separating its parts, than those of a block of the hardest stone. If a long foundation is required to be made, and the work is exposed to the violence of the waves, it is at first, only necessary to drive the piles of the *encaissement* following the intended lines, and only form the coffer with plank piling, in parts of six toises in length, to fill it immediately with masonry, that each part may resist the violence of a heavy sea, which would probably overturn an empty *encaissement*. In order to commence the work again after having continued the *encaissement*, the side partitions formed with plank piling are to be removed; if the *bèton* is already firm, some apertures must be made in the side, by which means the new masonry will key or bind better with that already erected. If, on the contrary, it has not become a uniform body, it will incline itself along the *encaissement*, and will naturally join with the *bèton* which is thrown over it.

It is almost needless to observe, that according to the circumstances of situation, &c. boats or lighters must be in attendance with proper rigging, to carry the materials and rafts of timber round the *encaissement* to assist in the work. The foundation having been left, at least during the winter, to allow it to harden, the masonry must be raised by regular courses; the facing should be of good hard and picked unhewn stone, whether for the construction of a mole, jetty, battery, or a fort. If the surface is exposed, and likely to be abraded by the action of the sea, or by continual concussions arising from external forces—as is the case in quays, &c.; it is imperatively necessary that the edging of stone should be of sufficient substance, the headers and stretchers placed "*en coupe*," joggled together with iron dowels run with lead.

If works of magnitude are to be erected, jutting far out and of considerable length—such as a jetty or a mole—it is sufficient to make an *encaissement* all round, to enclose the *bèton* masonry; then fill the space so enclosed with unhewn stone, after which continue to raise the same lining of good masonry in freestone, placed in regular courses up to the top of the platform, allowing a moderate size for the *encaissement*, so as not to create useless expense. In order that a greater degree of solidity may be allowed where necessary, the opposite walls should be bound by chain bars, placed at regular distances, effected either by *encaissement* or by *bèton* masonry. It may be added, that in building the circummuring of a fortification, or of any other work much exposed to the open sea, and when, notwithstanding the counter-jetty with which it is guarded, there is reason to fear that the waves might dash against the face of the work, causing much damage by their violence, the exposed works should be protected by covering them with inch-and-half or two-inch planking, securely bound together with cramps, and permanently fixed against the masonry for the first year, until the whole should have time to consolidate and harden, and that the joints should not open. As there are some remarkable particulars connected with the use of *bèton*, we shall probably interest those concerned in its use, by subjoining the results afforded by its extensive application at Toulon.

The statement contains the quantities of each kind of materials contained in a cubic toise of this masonry, and their weight. To fill a foundation containing sixteen cubic toises, is as follows :—

Cubic Feet.		lbs.
942	Of Red Pozzolano, at 90 lbs. the foot, making	84,780
471	Of Sand, at 115 lbs.	54,165
235	Pounded Iron Dross, at 80 lbs.	18,800
1,020	Rubble Stone, at 110 lbs.	112,200
706	Quick Lime, at 76 lbs. each	53,656
618	Stones, at 160 lbs.	98,880
3,992	Total cubic feet, weighing	422,481 lbs.

Thus it is seen that to make 16 cubic toises of *bèton* masonry, there is used 3,992 feet of materials, although these sixteen feet only contained 3,456 solid feet, thus the difference is 536 cubit feet, caused by the vacuity between the parts of the materials employed; respecting which it is to be remarked, that M. Milet de Montville has often proved that 2 cubic feet of rough sand, incorporated with 1 foot of slacked lime, would not produce more than 2 feet of mortar; from which it follows, that in the account of the materials employed in this masonry, care should be taken to make proper deduction for the lime.

CHAPTER IV.

ON THE ANALYSIS OF LIMESTONE.

THE extraordinary phenomena exhibited by the combination of lime with silicious matters, and the variable results that are effected by modifying the proportions of the component parts in the composition of cements and mortars, for building purposes, are of the utmost importance to the practical builder. An elucidation of the facts connected with this portion of our labours is of so much practical and scientific interest, that we shall offer no apology for taking, as far as our limits will permit, a comprehensive view of the subject. There are two theories relating to the formation of lime. The first of these was promulgated by the earlier geologists, and up to the present time, has been advocated by several scientific men. This theory contends for the origin of limestones from organized substances, or, in other words, that it may be "an animal product, combined by the powers of vitality from some simple elements."

The arguments deduced in opposition to this hypothesis have been ably pointed out by Mr. Lyell, in his "Elements of Geology," who contends that lime is formed by the evolution of carbonic acid, and a small quantity of carbonate of lime on the sites of springs. The immense deposits of carbonate of lime originating from springs in volcanic districts, he attributes, with great propriety of reasoning, to the solvent power of carbonic acid; and instances that, "as acidulous waters percolate calcareous strata, they take up a certain portion of lime, and carry it up to the surface, where, under diminished pressure in the atmosphere, it may be deposited, or, being absorbed by animals and vegetables, may be secreted by them."

Limestone,* when burned is deprived of its carbonic acid, &c. and becomes converted into *lime*, its chief constituents being a *metallic oxide*† combined with *oxygen.* For the manufacture of mortar, lime requires *slacking*, which is accomplished by wetting it with a certain proportion of water, a compound is thus formed, bearing the chemical name of "*hydrate of lime.*"

It is evident that, to acquire practical information concerning the properties of various limes, an analysis of a series of limestones must be the first step to obtain the requisite information. The student should, nevertheless, be apprised that chemical examination *alone* will not enable him to decide upon the quality of the material, but that his

* Limestones are compounds of *carbonic acid* with *salifiable bases.*

† The *metallic base* of lime is "*Calcium,*" and was discovered by Sir H. Davy, in the year 1808.

research should, in all cases, be accompanied by actual experiment. Calcareous minerals are commonly distinguished by their effervescing and dissolving in acid, and the readiness with which their surfaces may be cut or scratched with a knife; their *elements* may be readily and accurately ascertained by the analytical process hereafter described.

The analyses contained in this chapter have been mostly derived from the "Transactions of the Geological Society of London," a series of publications containing papers of great practical value to the engineer, &c. From the authenticity of those records, and the scientific eminence of the writers, it is anticipated that the various analyses thus collected and brought under immediate observation, will not fail to assist in developing the nature and quality of a most important class of building materials.

Lime may be obtained from an immense variety of stones; indeed, any stone exhibiting the indications mentioned above, will burn to lime; varying in quality in proportion as it is more or less impregnated with silex, alumina, iron, &c., the *pure limestones* furnishing the *weakest lime.** *Chalk* is the most common variety of calcareous mineral from which lime is obtained; its *physical* character is so well known as to need no remark. We shall merely observe, that when quarried the less exposure it

* M. Vicat states, "that no hydraulic mortar exists without silica, &c." An exception has, however, been observed by Captain Smith, in favour of the lime made from *Dolomitic* or *magnesian limestones* (see Captain Smith's translation of "*Vicat on Cements.*")

has received from the atmosphere, the better will be the quality of the lime. Most of the lime used in the neighbourhood of London and its suburbs is procured from Dorking, in Surrey. This lime is produced by burning masses of indurated chalk, or chalk marle. Chalk occupies an extensive and conspicuous geological feature in this island; its great abundance, and the ease with which it is obtained, renders it a most available medium for the manufacture of lime.

The specific gravity of an ordinary specimen of chalk is 2.3, and, according to the analysis of Bucholz, contains—

Lime	56 5
Carbonic Acid . . .	43 0
Water	0 5
	100 0

In its purest state it is a carbonate of lime; but the usual varieties contain much silicious matter, which is separated from it by pounding and washing.

Chalk may with propriety be considered as a "tender earthy limestone," the second in the scale of materials from which lime is made. Thus we consider lime made from the varieties of limestones, 1st; chalk, 2nd; and that made from shells, the 3rd. Near Dorking is one of the escarpments of the chalk range, and at two-thirds or three-fourths of the depth of the chalk strata, a very hard chalk is obtained, having a brownish tinge; this is quarried, broken, and burned, and forms the substance so well known by the name of Dorking lime. In proportion to the depth at which the chalk is obtained, so is its state of induration,—in this respect not differing from other strata. It has been remarked, that within a few yards of the bottom of the chalk formation, there are one or more beds of it so hard as to be nearly equal to the best Portland stone. In many places its induration is so great, that it has been quarried for the purposes of masonry; the following may be cited as instances of its employment as a building stone:—The abbey of Sturley, and its parish church, Berkshire; abbey of St. Omer's; mullions and arches of St. Catherine's chapel near Guilford, Surrey, &c.* A lime obtained from Merstham in the same county (Surrey), is also extensively used; it is obtained from an indurated chalk marle (clay and chalk), the kilns used for burning it into lime are distributed in various parts of the county; those at Merstham are situated at the lowest level at which the marle is fit for the kiln, beneath which it becomes much harder and partakes of the nature of stone.† The more argillaceous forms of chalk marle, according to the analysis of Mr. Phillips, indicate carbonate of lime nearly 30 per cent.; and the more cretaceous varieties 82 per cent. carbonate of lime, 18 silex and

* Conybeare.

† Middleton. Communication, Monthly Magazine, 1812.

alumine, chiefly the former, and a trace of the oxide of iron.

For the greater convenience of the reader, we shall follow (as closely as the subject will admit) the regular geological succession of strata, in order to exhibit successively the limestones peculiar to each formation. The limestone from which the greater number of the varieties of lime have been made and experimented upon, were of the sub-species.*

According to Dr. Ure, compact limestone common to the English mining districts is a carbonate of lime, having variable and minute portions of silica, alumina, iron, magnesia, and manganese; it effervesces with acids, and burns into quick lime. Sp. gr. 2.6 to 2.7.

Oolitic series. 1. Portland.—The summit of the Portland beds, called the *cap*, is only burned for lime. The cap is No. 4 of the layers of the quarries; it is a cream-coloured stone in three layers, with partings of clay, and so hard as to turn the steel points of chisels and pickaxes.

2. Cornbrash.—Forest marble, a loose rubbly limestone, affords lime of tolerable quality.

3. Bath stone.—Now extensively worked for the purposes of masonry: it is a tolerably pure carbonate of lime, but as a lime, possesses no remarkable features. The oolitic series contain such an

* For this enumeration, see Ure's Chemical Dictionary. Article "*Limestone.*"

immense number of intermediate beds of stone, capable of being converted into lime of variable quality, that the bare enumeration of them would occupy a very considerable space. It is sufficient for our present purpose to observe, that throughout the whole tract of country occupied by these formations,* little difficulty is experienced in procuring limestone fit for the purposes of the arts, particularly that of masonry. They are, in fact, the grand depositories from whence we draw the materials for our supply of building-stone; and furnish, besides, inexhaustible materials for the use of the brickmaker and lime-burner.†

The series alluded to above (geologically speaking) include all the strata between the iron sand and the red marle. These comprise three great divisions, called the upper, the middle, and the lower oolitic systems or divisions; each separated in a remarkable manner by an extensive argillaceous district or valley. This important circumstance so well defines the boundary of each division, that the architect or engineer may readily detect the general quality of the subsoil by the locality of the separations. The stony beds of the first division are the Purbeck, Portland, Tisbury, and Aylesbury limestones, &c. Then follows the valley of clay, known generally as the oak-

* Comprising a zone having nearly thirty miles average breadth, and extending across the island from Yorkshire, on the north-east, to Dorsetshire, on the south-west.—(Conybeare).

† The chapter on building-stone will describe the more remarkable divisions of the strata.

tree clay. 2nd. Coral rag, &c. &c. divided by the Oxford or clunch clay. 3rd. Corn-brash, or corn-grit, forest marble, great and inferior oolite, containing the Bath stone, &c. The lias, and lias marle forming the base of the whole.

For the manufacture of cement, the lias limestone must be allowed to hold the most prominent place in the above series. In the neighbourhood of Bath, the dislocations and disruptions of the inferior oolite, together with quarrying through the strata of stone, have laid open the lias formation; from hence is supplied the lias lime, so extensively used in Wiltshire and the adjacent districts. In the recent alterations at Heywood House,* at Westbury, and at other extensive works in Wiltshire, the lias lime has been used as a cement in all underground or damp situations: its known property of setting under water renders it invaluable for such

* In Wiltshire and Somersetshire, the provincial terms for the beds of stone available for building, and procured from the Box-stone-quarries, are, 1, Corn-grit; 2, Scalit; 3, Ground stone; 4, Farley-down stone; and 5, Coombe-down stone. The corn-grit and scalit are of a tolerably fine texture, but not waterproof. Ground stone is an excellent weather stone, though somewhat coarse: it is well adapted for plinths, quoins, and projections that are much exposed to the weather. Farley-down stone ought to be employed for carvings and mouldings in interior work, but is not proof against the weather; if employed externally, sufficient time should always be allowed for the rock water (which it contains) to evaporate, otherwise the first attack of frost will disintegrate the whole mass, causing it rapidly to fall to pieces: it should not be used in damp situations. Coombe-down stone is of a coarser quality than Farley-down stone, it is much harder, and of a gray colour: it is generally considered by masons as weatherproof, and may be used for almost any external purpose: the top beds are considered the best.

purposes. In the immediate neighbourhood, it is known among masons by the name of Bath brown lime, and when prepared for cementing, or in combination with the patent metallic cement, is what is locally termed "*wind slacked;*" namely—after having been burned, it is placed in covered sheds, but open at the sides, the atmosphere being allowed to operate upon it; should the slacking proceed too slowly, a small quantity of water may be sprinkled upon it to stimulate the process, but on no account should water in a considerable quantity be added; it is therefore much better (if possible) to allow the atmosphere to act for this purpose. The lime, when thus slacked, is converted into fine granulated particles, and is among workmen said to be "alive," as it will run from an iron shovel similar to quicksilver. The colour of the lias previous to burning, is blue; when it has passed the kiln, it is brown. The uniformity of its mineralogical character is well retained throughout, including also its foreign localities, which are very extensive: a correct knowledge of the lias limestone is, therefore, of almost universal application. Although a minute analysis of the lias limestone has not yet been made, we may calculate upon the greater number of specimens containing as much as 90 per cent. of carbonate of lime, the residuum in all probability consisting of alumen and iron. Mr. Smeaton experimented upon the following varieties of lias lime, by dissolving 40 grains of each in aquafortis, and obtaining a precipitate from each amounting to the following quantities:—

		Grains.
Yellow lias, of Axminster	. . . .	5¾
Do. with shining spangles	. .	5½
Yellow such-stone, of Glastonbury	.	5
Blue lias, of Watchet	. . . . . .	4½
Do. Aberthaw (Wales)	. .	4½
Do. Bath	. . . .	4½
Do. Axminster*	. . . . .	3½

The next important limestone deposit† is that associated with the red marle or new red sandstone. This is denominated the newer magnesian, or conglomerate limestone,‡ from its containing an excess of magnesian earth. The chemical and external characters of this limestone, are a granular sandy structure, glimmering lustre, and a yellow colour; it contains about 20 per cent. of magnesia. The abundance of this limestone in England renders it an important object of notice, and the analysis of its varieties will not fail to furnish useful information to the engineer. It should be observed, however, that in the composition of rocks, (and more particularly the one in question), that the same anomalies present themselves that we find occurring so repeatedly in experiments upon the strength and stress of materials, where it has been found, that

* Conybeare, Geol. Eng. and Wales, p. 263.

† "The limestone of St. Vincent's rocks near Bristol, (associated with the lias) when calcined, yields a very pure lime; large quantities of it are exported for the use of the sugar works in the West Indies, in an unslacked state, and packed in tight casks, it is used extensively in building."—Dr. Bright on the strata, in the neighbourhood of Bristol, vol. 4, Geol. Trans. p. 200.

‡ Also called dolomite.

scantlings of the same size, from the same mass, and apparently of the same quality, offer results widely differing from each other—so in the magnesian limestone formations: in one place we meet with a common limestone, and in another spot not far distant an excess of magnesia; even the same stratum (observes Dr. Gilley) "will vary in its colour, hardness, and general structure in different parts of its course." With the *rationale* of this phenomena we have, at present, nothing to do; the fact is here mentioned with a view to guard the enquirer against the apparent contradiction that may arise in the course of his investigations. The principal range and extent of the magnesian conglomerate formation is from Sunderland to Nottingham. It makes its appearance in various other parts of this island, as will be seen by the localities described in treating of the several specimens selected for analysis. The continuity, however, is most apparent through the line of country alluded to above. A specimen of the magnesian limestone of Yorkshire, has been analysed by Smithson Tennant, Esq.; it was procured from the stone of which York Minster is built, and found to contain as follows:—

Carbonic acid . .	47	00
Lime	33	24
Magnesia . . .	19	36
Iron and Clay . .	0	40
	100	0

A similar variety, namely :—that of Westminster Hall yielded results nearly alike, but exhibiting about two per cent. less of magnesia.

The magnesian limestone of Sunderland is situated to the north-west of the red sandstone. The northern extremities of the western boundary of the magnesian limestone is exposed in the cliffs at Cullercoats in Northumberland. On the coast in the vicinity of South Shields, in the county of Durham, the formation becomes extensive, and may be followed till it reaches the Tees below Winston-bridge, thirty miles south-west of that river's junction with the sea, and for four miles from the Tyne at South Shields. The celebrated quarry at Whitby, near Cullercoats, offers an admirable opportunity for the examination of the limestone of this district, and is minutely described by N. J. Winch, Esq., (4th vol. of the Geological Transactions,) to whom we are indebted for a highly valuable paper, on the geology of Northumberland and Durham, from which the following account is extracted :—" A hollow space, formed like a basin or trough, is filled with the limestone; the length of this from east to west is about a mile; the breadth from north to south four hundred yards; the depth seventy feet. The beds pass over the Ninety-fathom Dike, which has occasioned in them no confusion or dislocation; so that there can be little hazard in stating, that the beds of the magnesian limestone belong to a more recent formation than those of the Coal-field.

The limestone has been quarried across its whole breadth, and a numerous set of thin strata are thus exhibited to view. At the surface loose blocks of a bluish grey coralloid limestone, the produce of the lead mine district, are found imbedded in the soil. Three or four of the uppermost strata of the quarry are of white slaty limestone, which being nearly free from iron, burns into a pure white lime. Below these an ash-gray firm grained stratum is met with, which strongly resembles a sand-stone, and seems to contain nearly as much iron as the ferri-calcite of Kirwan, becoming magnetic by the action of the blow-pipe: it produces a brownish yellow lime, less esteemed for agricultural purposes than the former. The beds next in succession are of an ash-gray colour, are compact in texture, and conchoidal in fracture; these afford a buff-coloured lime, which sells for nearly the same price as the white: near the bottom of the quarry the limestone alternates with shale; the whole rests upon a stratum of shale on the southern side, and upon a thick bed of sand-stone on the northern. The shale has been cut through to a considerable distance from the kilns in the direction of North Shields, for the purpose of laying a railway to the Tyne. The thickness of the limestone strata varies from three or four inches to as many feet. Small strings of galena have been found here, and in one of the strata, that was walled up when I visited the quarry, a few organic remains have been noticed. The stone intended to be burned is detached from the rock by the agency of fire, during which process those portions which contain iron become of a brick-red colour. Considerable quantities of fuel are found necessary at the kiln, and some parts of the rock are too apt to vitrify in the process; an accident to which the crystalline limestone of Sunderland is not liable."

The limestone quarried in the vicinity of Sunderland is of a brown colour; it contains inflammable matter, and, consequently, requires less fuel for its conversion into lime.* The Northumberland and Durham limes exhibit the following products from an analysis of 100 parts of each:—

By the Rev. J. Holme, from Denton, near the Tees.

Carbonate of Lime	63	0
Ditto of Magnesia	30	0
Alumina, Red Oxide of Iron and Bitumen	2	25
Water	0	25
	100	0

By Sir H. Davy, from Eldon.

Carbonate of Lime	52	0
Ditto of Magnesia	45	2
Iron	1	1
Residuum	1	7
	100	0

* The exportation of lime from Sunderland is chiefly to Scotland, and amounts to from forty-two to forty-five thousand chaldrons, of thirty bushels each, annually.

By Sir H. Davy, from Aycliff.

Carbonate of Lime	45	9
Ditto of Magnesia	44	6
Iron	1	57
Residuum	2	8
	100	0

Some specimens of this lime have the property of setting under water; and when mixed with bullock's blood forms a strong iron cement, occasionally used for repairing boilers. The lime from Leigh and Ardwick was employed at Drury-lane Theatre, and used as a Tarras in those works. In the neighbourhood of Bristol there is a remarkable instance of the conglomerate limestone appearing in contact with the new red sandstone. The basis of the red ground, or new red sandstone, is in most cases a common limestone; but at the village of Portishead, near Bristol, a discovery was made by Dr. Gilby, of a limestone containing a considerable quantity of magnesia; the stone was of a yellow colour, resembling the Yorkshire magnesian limestone. In some strata it is so mixed with sand as to give more than 20 per cent. of insoluble matter. The fragmented portions are generally limestone or red sandstone; but it is found that other strata are destitute of sand and fragments, forming in fact a hard compact magnesian limestone. This variety will give 36 or 38 per cent. of carbonate of magnesia. A compact variety yields as follows :—

Carbonate of Lime	53	5
Ditto of Magnesia	37	5
Oxyde of Iron	0	8
Insoluble matter	7	0
Loss	1	2
	100	0

A small ridge of rock, about four miles north-west of Bristol, upon which Lord de Clifford's house is built, is entirely composed of a magnesian limestone. A specimen yields as follows :—

Carbonate of Lime	58	0
Ditto of Magnesia	38	0
Oxyde of Iron	1	1
Silica and Bituminous matter	1	5
Loss	1	4
	100	0*

The quantity of magnesia contained in limestone does not appear to operate injuriously upon lime employed for building purposes, although its presence, when used for agricultural purposes, is anything but favourable to vegetation. The presence of magnesia in limestone used for building, produces scarcely any other effect than to render the proportion of clay greater in relation to the

* Dr. Gilby on the magnesian limestone in the neighbourhood of Bristol. Vol. iv., Geol. Trans., p. 212.

proportion of carbonate of lime. In order to appreciate the qualities of a calcareous stone as a limestone, it is sufficient to determine the quantity of argile and magnesia contained in it. The simple process by which this is effected is thus described by M. Berthier in the *Annales de Chimie.** "Pulverise the stone and pass the powder through a silken sieve; put ten grains of this powder into a capsule, and pour over it, a little at a time, muriatic acid, nitric acid, or vinegar diluted with a small quantity of water, agitating it at the same time with a glass tube or small stick; discontinue adding the acid when it ceases to effervesce; then evaporate the solution with a gentle heat, until the whole is reduced to a pasty consistence; dilute it in about half a *litre* of water, and filter it; the clay remains on the filter, and must be dried in the sun or by a fire, and weighed; or it will be better before weighing to calcine it to a red heat in an earthen or metal crucible. Pour very limpid lime-water into the solution, as much as to cause a precipitate; collect this precipitate, which is the magnesia, as quickly as possible upon a filter; wash it with pure water, calcine or otherwise dry it as much as possible, and then weigh it. If there be any iron or manganese in the solution, they will precipitate with the magnesia. It would be superfluous to attempt to separate these three substances from each other."

* See also Repertory of Patent Inventions, vol. 43, N.S. p. 185-6.

Carboniferous Limestone.—Associated with the all-important coal districts of this country is a vast geological formation known by the above name, which significantly points out its characteristic locality, and at the same time serves materially to mark the distinction between the magnesian limestone beds associated with the red ground or new red sand-stone. This rock exhibits generally a close and compact body, capable of a good polish, and imperfectly crystalline,* prevailing colour gray. The purest beds contain about 96 per cent. of calcareous matter. By admixtures, not common to the pure species, it passes into magnesian, ferruginous, bituminous, and fœtid limestone. For the manufacture of lime, possessing all the essential qualities of a cement, capable of the highest degree of induration, the carboniferous limestone of England is upon the whole the most important in the class of calcareous minerals. From some unaccountable reason, the valuable properties of this mineral have not been sufficiently appreciated by the profession, its use has, consequently, been very limited in our metropolitan works. In engineering operations good lime is a material that ought to be obtained, regardless of expense. Indeed, when it is considered, that artificial hydraulic limes, such as are generally used in the metropolis, are obtained from a variety of ingredients collected in various districts

* Conybeare

distant from each other, and subjected to careful manipulation, it becomes a question whether the conveyance of the best varieties of this mineral to the London market, would not prove of sufficient practical value to counterbalance the excess of outlay required in its general employment in our public works. The carboniferous limestone of Flintshire, North Wales, (parish of Whiteford) exhibits in an eminent degree the valuable properties of the bituminous limestone. A careful analysis of a variety, consisting of 100 parts, was made at the request of Mr. Pennant, by Edward Daniel Clarke, L.L.D. Professor of Mineralogy in the University of Cambridge, and published in the fourth volume of *The Geological Transactions.* We subjoin the process adopted, together with the results: these afford an admirable example of chemical investigation, adapted to the examination of that description of minerals, which forms so prominent a class in the economy of engineering. The specific gravity of the limestone, (alluded to above), estimated in pump-water at a temperature of 50° Fareinheit = 2,670. It is of a dark brown colour, and, when breathed upon,* exhales an earthy odour, denoting the presence of *iron* oxide, in combination with *alumine;* but its colour is owing to *bitumen* rather than to *iron.*"

1. One hundred grains being placed upon red

* All the shaly limestones may be tested by this method. They emit that peculiar earthy smell, common to the hearths of fire-places when rubbed with a wet hearth-stone in a warm state.

hot iron for the expulsion of the *water of absorption*, were thereby diminished $\frac{5}{20}$ of a grain.

2. The remainder being reduced to powder in a porcelain mortar, and exposed to diluted muriatic acid until all effervescence ceased, there remained an insoluble residue of the original dark colour of the limestone, which, when carefully washed and dried, weighed ten grains; allowing, therefore, for the weight of the *carbonic acid* and *lime*, after the expulsion of the water of absorption, $89\frac{1}{4}$ grains.

3. The supernatant acid used in this experiment being decanted, and neutralized by the addition of an alkali, yielded no precipitate of iron to the tincture of galls; but the prussiate of potass threw down a blue precipitate, upon which, however, no reliance can be placed, as it is well known that the prussiate of potass is not a satisfactory test of the presence of *iron* when this metal exists in an inconsiderable portion.

4. The ten grains of dark brown powder, mentioned in No. 2, being collected, washed, and dried, were exposed to the heat of a flame of a candle, urged by the common blow-pipe, when combustion instantly ensued, accompanied by a lambent flame, which continued during some seconds, the powder thereby losing its colour and becoming white; attended also by a loss of weight, amounting to $\frac{2}{5}$ of a grain. Hence it is manifest that the colour is owing to *bitumen*.

5. To ascertain the proportion of alumine,

(which, from its chemical combination with silex, remained insoluble in the muriatic acid,) a plan recommended by Mr. Holme was adopted. One hundred other grains of the same limestone were calcined in a platinum crucible, and the loss of weight, owing to the expulsion of the *carbonic acid* was found to equal $40\frac{1}{10}$ grains.

6. The calcined residue being placed in muriatic acid, a solution now took place both of the *lime* and the *alumine*, and there remained at the bottom of the vessel only an insoluble portion of pure silex, in the form of a white powder, which, when carefully washed and dried, weighed $\frac{3}{5}$ of a grain. Deducting, therefore, this weight of the *silex* from the weight of the *silex* and *alumine*, which remained in No. 4 after the combustion of the bitumen, the weight of the *alumine* is ascertained, which, of course, $= 8\frac{8}{10}$ grains."

Thus we have, by the foregoing analysis, as follows :—

Lime	49	65
Carbonic Acid	40	10
Alumine	8	80
Silex	0	60
Bitumen	0	60
Water	0	25
	100	0

The valuable property of the mortar prepared from this *limestone* is owing to the presence and proportion of *alumine*; and to its property of rapidly absorbing water. It may be added, that the cement possesses all the properties of the *pulvis puteolanus* of the Romans.* The following chapter contains observations on the nature and properties of other specimens of British limestone. With respect to the limestone of the *continent*, the admirable translation of M. Vicat's work, by Capt. Smith, and some papers by M. Berthier, translated, and inserted in the 43rd Vol., N. S., of the *Repertory of Patent Inventions*, are references to which the student may readily avail himself with considerable advantage. We have now placed before the reader the analysis of the grand divisions of the limestone formations of this country. Each subdivision of the limestone strata may, however, be found to vary in its component parts, and in its relative situation. In the choice of stone, and in its application, the analysis already given will materially assist any further investigation; for the rest, observation and experience must decide.

* An analysis of this substance will be found in the following chapter.

CHAPTER V.

ON THE CALCINATION OF LIMESTONE.—OBSERVATIONS ON BRITISH LIMESTONES, &c.

LIME has not hitherto been found in a pure or native state ; but in combination with acids and earths it exists in vast quantities. In order to free it, as much as possible, from carbonic acid gas, water, &c., and to render it fit for the purposes of the arts, the limestone must undergo the process of calcination, or burning. The effect of various degrees of calcination upon limestone, has been carefully noted by Dr. Bry Higgins,* who, in the year 1775, instituted a series of experiments for this purpose. Having ascertained, from the experiments and researches of Dr. Black, that calcareous stones, which burn to lime, contain a considerable quantity of acidulous gas, and that this gas, in chemical combination with earthy matter, formed the greater

* "Higgins on Calcareous Cements," 8vo. 1780. Since the publication of this work, the progress of chemical science has been so rapid, and the results of later discoveries have elicited so many new and important facts, that the philosophical deductions of the writer have been thereby rendered obsolete. It must be admitted, however, that the scientific *data* obtained from several of his experiments are of the highest value; among these may be instanced the observations (No. 1 to 17) on the calcination of limestone, given in the text.

portion of the mineral known as limestone, he (Dr. Higgins) proceeded to make the following experiments, upon a scale sufficiently extensive to attest the value of the results.

"It should be observed, that the difference between chalk lime and the lime obtained from the various limestones, chiefly consists in the greater retention or expulsion of the carbonic acid gas contained in them. Specimens of different limestones and chalk were procured, and broken into small pieces; these were burned in close crucibles, lined with lime, to prevent adhesion and vitrification. Similar specimens and quantities were also operated upon in perforated crucibles, thereby admitting a free current of air through them during the firing process. Similar specimens, of three pounds in weight each, were also subjected to a distillating process over a graduated fire, in an earthen retort, of a size barely sufficient to hold this quantity. The neck of each retort was lengthened with a conical glass tube, luted on with four parts of lime, one of fine sand, and as much common glue as was sufficient to form a paste. This luting was found impervious, and in every way sufficient for the success of the experiments. The extremity of the glass tube was immersed in mercury, and by inverting a bottle filled with mercury over the extremity of the tube, whatever water or gas was expelled from the calcareous matter by the action of the fire was carefully collected and measured. The following observations were deduced from the foregoing processes :—

"1. Limestone or chalk, heated to redness only, in a covered crucible, or in a perforated crucible, through which the air circulates freely, loses about one-fourth of its weight, however long this heat be continued. The sort of lime so formed effervesces considerably in acids, slakes slowly and partially to a powder, which is not white, but is gray or brown, and heats but little in slaking.

"2. Limestone or chalk, exposed to a heat barely sufficient to melt copper, whether in a perforated crucible or otherwise, loses about one-third of its weight in twelve hours, and very little more in any longer time. This lime effervesces but slightly in acids, it heats much sooner and more strongly than the foregoing when water is sprinkled on it, and it slakes more equally, and to a whiter powder. In a variety of trials this lime appeared to be in the same state with the best pieces of lime prepared in the common lime-kilns: for the quantities of acidulous gas obtainable from both by a stronger heat, or in solution, were nearly equal. They slaked in equal times, with the same phenomena, and to the same colour and condition of the powder.

"3. Lime burned in perforated crucibles, or in the naked fire, is whiter than that burned in common crucibles; in which case the air has not so free access to it, although the loss of weight be the same in both; but this latter kind of lime, in slaking, affords as white a powder as any other which has lost equally of its weight.

"4. When dry chalk or limestone is used, in the process above described for making lime in close vessels, and for examining the matter which is expelled by fire, the quantity of water obtainable from it by heat is so inconsiderable as to deserve no notice in our mensuration of that matter.

"5. Chalk or limestone heated gradually in these close vessels, lose very little acidulous gas until it begins to redden; after this the elastic fluid issues from it the quicker as the heat is made greater, and continues to issue until the retort glows with a vivid white heat, sufficient to melt steel.

"6. Forty-eight ounces of chalk yield twenty-one ounces of elastic fluid, the first portions of which are turbid as they issue, but soon become clear, without loss of bulk, by the condensation of the watery vapour; the remaining portions issue transparent and invisible.

"7. There siduary lime of forty-eight ounces of chalk, urged with such heat to the total expulsion of the elastic fluids, weighs only twenty-seven ounces, whilst it is red hot. When it cools it weighs more by reason of the air, which it imbibes as the fire escapes from it.

"8. When no more heat is employed than is necessary for the expulsion of these elastic fluids, the residuary matter is found contracted sensibly in volume, and is good lime, although not so white as lime prepared in the usual way. With water it slakes instantly, grows hissing hot, and perfectly white. The slaked powder is exceedingly fine, except in those parts of the lime which lay in contact with the retort, which are always superficially vitrified, because clay and lime promote the vitrification of each other.

"9. The lumps of this lime, immersed in lime water, or boiling water, to expel the air which such spongy bodies imbibe in cooling, dissolve in muriatic acid without shewing any sign of effervescence.

"10. Limestone or chalk, gradually heated in a crucible, or in the bed of a reverberatory furnace, or in contact with the fuel in a wind-furnace, does not become perfectly non-effervescent and similar to the lime last described, in slaking instantly, and growing hissing hot when water is sprinkled on it, until it has, after a strong red heat of six or eight hours, sustained a white heat for an hour or more.

"11. Limestones heated sufficiently to reduce them to lime, which slakes instantly, with the signs above described, and which is perfectly non-effervescent, do not, in general, lose so much of their weight as chalkstone does, under similar treatment. Some limestones lose little more than the third of their weight. Those which lose the most, slake the quickest, and to the finest powder; and those which lose the least, slake the slowest, and to a gritty powder, composed of true lime and particles, chiefly gypseous.

" 12. The quantity of gypsum, or of other earthy matter, in well burned lime, is discoverable in dilute muriatic acid; for this dissolves and washes away the lime, leaving the gypsum to be measured when dry, the part of the gypsum which dissolves being too small to deserve any attention. And if any other earthy matter, or any saline matter existed in the limestone, it vitrifies with part of the calcareous matter in the heat necessary for making non-effervescent lime, and is separable by the means last mentioned, and even by a fine sieve, in most instances.

" 13. When limestone or chalk is suddenly heated to the highest degree above described, or a little more, it vitrifies in the parts which touch the fire vessels, furnace, or fuel, and the whole of it becomes incapable of slaking freely, or acting like lime. Limestone is the more apt to vitrify in such circumstances, as it contains more gypseous or argillaceous particles, and oyster-shells or cockle-shells vitrify more easily than limestone or chalk, when they are suddenly heated, which may be imputed to their saline matter; for when they are long weathered they do not vitrify so easily.

" 14. The agency of air is no further necessary in the preparation of lime, than as it operates in the combustion of the fuel.

" 15. Calcareous stones acquire the properties of

lime in the most eminent degree, when they are slowly heated in small fragments, until they appear to glow with a white heat, when this is continued until they become non-effervescent, but is not augmented. The art of preparing good lime consists chiefly in attending to these particulars.

"16. That lime is to be accounted the purest and fittest for experiment, whether it be the best for mortar or not, which slakes the quickest, and heats the most in slaking; which is finest and whitest when slaked; which, when wetted with lime-water, dissolves in muriatic acid or distilled vinegar without effervescence, and leaves the smallest quantity of residuary undissolved matter.

"17. The quick slaking, the colour of the slaked powder, and the former acid, are the most convenient, and, perhaps, the best tests, of the purity of lime. The whiteness denotes the lime to be freed from metallic impregnation, the others shew any imperfections in the process of burning, and the heterogeneous matter inseparable from the calcareous earth by burning."

The results of the foregoing experiments are, perhaps, of greater importance to the scientific experimentalist than to the practical lime burner. The calcination of limestone *in small quantities* may be effected with tolerable success and uniformity of operation in the laboratory; but, on a comparison with the effects produced by the more extended operations of *kiln-burnings*, certain anomalies present themselves, which (as yet) have not been satisfactorily explained. Before adverting more particularly to these, we proceed with a description of the various kilns (suited to particular districts), and best adapted for the calcination of limestone on an *extensive scale.*

Several attempts have been made to economise the calcination of limestone by the preparation or manufacture of other materials that required the agency of heat. We may particularly instance the manufacture of *coke*, and the operation of *brick-burning* in conjunction with *lime-burning.*

The first of these contrivances—namely, "a combination of *coke-ovens* with a *lime-kiln*," was the subject of a patent granted several years ago to Mr. Heathorn, of Maidstone. This gentleman had, for upwards of ten years, been employed in examining and experimenting upon a variety of kilns, and eventually succeeded in constructing a kiln that should prepare coke, at the same time that the limestone underwent the process of calcination.

By this process the whole of the fuel employed is saved by its being converted into coke; in addition to this, the *drawing* of the lime may be accomplished with a facility equal to that of a common tunnel-kiln; the chalk or stone is also calcined so gradually that no refuse is left, and consequently the operation of *sifting* is dispensed with. Kilns on this construction are in operation at Maidstone, Hackney, and at Hastings.

The following description of the kilns and the mode of working is derived from an account furnished by the patentee, and inserted in the "Register of Arts and Journal of Patent Inventions."*

"The drawing, Fig. 1, Plate 10, represents a vertical section of the lime shaft and coke-ovens, A A are the side walls (about four feet in thickness), of a rectangular tower, the internal space being filled with limestone from the top to the iron bars, B B at bottom, whereon the whole column rests. The limestone is raised in a box D, or other proper receptacle, to the top of the building, by means of a jib and crane, E, or other tackle, which is fixed at the back of the tower, together with a platform projecting from beyond the walls, for affording security and convenience for landing. When the limestone is raised as represented, the jib is swung round, and the lime box tilted, by which the whole contents are thrown into the shaft.

"The coke-ovens, of which there may be two, or a greater or lesser number, according to the magnitude of the works, are constructed and arranged in connexion with the lime shaft, in the same manner as the two represented in the drawing at F F. These ovens are supplied with coal through iron doors in the front wall (not seen in the section), the doors have a long and narrow horizontal opening in the upper part of them to admit sufficient atmos-

* No. 91, Vol. iv., p. 290-91.

pheric air, to cause the combustion of the bituminous or inflammable part of the coal; the flames proceeding from thence pass into the lime shaft through a series of lateral flues, (two of which are brought into view at G G), and the draft is prevented from deranging the process in the opposite oven by the interposition of the partition wall, H, which directs the course of the heat and flames throughout the whole mass of the lime; the lowermost and principal portion of which attains a white heat, the upper a red heat, and the intervening portions, the intermediate gradations of temperature.

"When the kiln is completely charged with lime, the openings in front or beneath the iron bars, at I I, are closed and barricaded by bricks, and an iron cased door, which is internally filled with sand, to effectually exclude the air, and to prevent the loss of heat by radiation.

"Therefore, when the kiln is at work, no atmospheric air is admitted but through the narrow apertures before mentioned in the coke-oven doors. When the calcination of the lime is completed, the barricades at I I are removed, the iron bars at B B are drawn out, by which the lime falls down and is taken out by barrows. It sometimes happens, however, that the lime does not readily fall, having caked or arched itself over the area which incloses it, in which case a hooked iron rod is employed to bring it down. To facilitate this operation in every part of the shaft, where it may be necessary, a series of five or six apertures, closed by iron doors, are made at convenient distances from the top to near the bottom of the shaft, two of these are brought into view at K K, where their utility is made apparent in the drawing. Two similar apertures are shown in section at the coke-ovens at L L, which are for the convenience of stoking or clearing out the lateral flues, G G, from any matter that might obstruct the free passage of the flames and heated air.

"When the coals have been reduced to coke, the oven doors in front (not shown) are opened, and the coke taken out by the peel iron, the long handle of which is supported by a swinging jib, that acts as a moveable fulcrum to the lever or handle of the peel, and thus facilitates the labour of taking out the contents of the oven."

The combination of a *brick-kiln* with a *lime-kiln* is suited to particular districts, where coals are scarce, or where economy is much studied. The combustibles employed are faggots or furze, &c.; and when limestone is thus calcined, it is technically termed "flare" or "flame burnt."

The construction of these kilns is as follows:—If they are built independent of any other support, the walls should be from four to five feet in thickness, and the face of the work made to batter upwards. In external appearance, they are in the form of a truncated pyramid, whose base externally measures between twenty and thirty feet, and the altitude about fifteen feet. The furnaces of the kiln consist of several arched openings (the number varying with the size of the kiln), these openings extend from the front to the rear, and are arched throughout with large loose pieces of the limestone intended to be burnt. Upon the crown of the arches the limestone is now heaped up to the height of eight or more feet, and upon the top are laid from fifteen to twenty thousand of new unburnt bricks. When the kiln is about to be fired, small quantities of fuel are placed under the arches, and the *external* openings are then bricked up nearly to the top—sufficient room only being left to add to the fuel as occasion may require.

A kiln of the above form and dimensions will calcine the whole of the limestone (about 120 quarters), and burn the bricks in the space of forty hours.*

All permanent or perpetual draw-kilns should be well dried by fire, previously to their being employed for burning the lime. The great heat to which they are subjected renders this precaution necessary to prevent their premature destruction by expansion.†

The practice adopted in some parts of Scotland for the calcination of limestone, is by means of a

* See Wild's Instructions to Emigrants. A work containing much valuable information on the subject of economical building.

† A recent examination of a furnace for reducing iron ore, at Stavely, near Derby, shewed an expansion of *nearly eighteen inches.* The furnace is built of brick, and similar to a lime-kiln in form, the diameter at the widest part outside is about seventeen feet.

kiln constructed according to the drawing Fig. 2, Plate 10. This kiln is a modification of one called "Booker's kiln," and is considered by competent judges, to be superior to most others where coal or other smoky fuel is employed. A particular account of the kiln has appeared in the *Gardeners' Magazine*, from the communication of C. J. S. Monteath, Esq., a gentleman who has had extensive experience in the quarrying and burning of lime.

The following remarks, extracted from Mr. Monteath's paper, apply more particularly to the arrangements and proportions to be attended to in its construction:—

"The kiln is built in a similar situation to 'Booker's kiln' (on the side of a bank); is of an oval form, and thirty-five feet in height; the diameter at the bottom, next the fuel chamber, is only twenty-two inches; but gradually extends, till, at the height of twenty feet, it is five feet; which dimension is continued to the top, where the oval (*ellipse*) is nine feet by five. There is an arched cover to the top (as represented in the drawing), mounted on small wheels, and is drawn off and on by windlasses H H, and has two small openings I I, serving as chimneys for the exit of the smoke.

"As the fuel chamber to this kiln is very broad in proportion to its depth, three separate doors or openings are found necessary, as well as advantageous, for more speedily and easily drawing out the lime. In some cases, instead of a moveable cover, a permanent roof of masonry may be adopted, which should have proper openings to admit the supply of lime and fuel (closed by sliding shutters or hinged doors), while on the roof there should be a chimney for the escape of the smoke.

"The chief use of a cover, whether fixed or moveable, is, of course, to retain the heat; but where it is a fixed structure, and sufficiently large, something will be gained by placing the fuel and limestones there, to be dried and heated before they are thrown into the kiln. Three-fifths of the '*Closeburn oval kiln*' may be drawn out every day; and when it is closed at top and bottom, the fire will not go out for five or six days."

Yorkshire Lime Burning.

Yorkshire abounds in a variety of limestones; some of them being particularly fitted for the purposes of agriculture, (such are those obtained from the southern edge of the eastern moorlands), and others no less excellent for building purposes, namely, the limestones of Brodsworth,* Roach Abbey, and subjacent places.

In a country possessing such an abundant supply of this mineral, and where its use and exporta-

* The prices of Brodsworth and Roach Abbey stone (1835) are for the first, 11d per foot (cube), and Roach Abbey 13d per foot (cube), each shipped at Doncaster.

tion are so extensive, it may be readily supposed that efficient means for its calcination have not been forgotten. The drawings, figs. 3, 4, 5, plate 10, exhibit the construction of the Yorkshire lime-kilns. The fuel used in these kilns consists of small coal interstratified with the limestone.

A correspondent in the *Mechanics' Magazine* a resident in the county, (vol. vii., p. 178), observes, that in fixing on a place to build such a kiln, the side of a hill, near the rock from which the limestone is obtained, should be chosen. The operations of the workmen are commenced by excavating a sufficient space to receive the back of the kiln; in erecting the shaft, two walls are to be carried up with a space between them. The vacuity is to be filled with "small rubble" (or sand) in order to prevent the radiation of heat, and the inner wall must be faced with a grit stone facing, about a foot or eighteen inches wide. This stone will stand an intense heat; when repairs are required, the wall in the rear prevents the materials from falling in.

The cost of the kiln, built according to the drawings, will not exceed 30*l.* Fig. 3, is an elevation of the kiln built against the limestone rock. Fig. 4, plan of ditto, and fig. 5 represents a section through the centre of the kiln. The various detailed dimensions are annexed to the drawings.

The reader will observe, that the construction of the above kiln is similar to the furnaces employed for reducing iron ore, in some parts of Derbyshire.

In certain districts of Yorkshire, Shropshire, Wales, and Scotland, where lime is required for manure, a singular practice prevails of calcining the limestone *without the use of kilns.* The limestone is calcined according to the method usually adopted for preparing charcoal. The stone is taken in *large* pieces, placed in a cavity in the ground, and heaped up in a conical form (called *coaks*) and to retain the heat and flame, is covered over with sods and earth.

Dorking has been long celebrated for the quality of its lime. It has been for many years extensively employed in and about London for works, which at least have acquired the reputation of being constructed with approved materials.

One of the most extensive establishments for quarrying and burning this lime, is that of Messrs. Bothwell, situated at Dorking, on the spot where some of the best rock chalk is to be obtained. The construction of the kilns differ materially from those already described (see plate 10). Through the kindness of the proprietors, we are enabled to give the drawings of one of their kilns, (see plate 11,) which is of excellent construction, and found to answer the purpose in the most efficient manner.

Fig. 1 (plate 11) represents the elevation of the kiln. Fig. 2 is a longitudinal section of the same, (taken through the *throat and furnace,*) and fig. 3 represents the ground plan. The several dimensions and references are subjoined, in order to render the construction more intelligible and available to the engineer.

The limestone of Ireland is, for the greater part, crystalline—much of it, in fact, being marble, of excellent quality. The form of the Irish *permanent kilns* is generally that of the "egg placed upon its narrower end, having part of its broader end cut off." The compact texture of the limestone requiring considerable heat for its calcination, this form has been adopted, and is said to possess the following advantages:—

1st. "The upper part of the kiln being contracted, the heat does not fly off so freely as it does out of a spreading cone.

2nd. "That, when the cooled lime is drawn out at the bottom of the furnace, the ignited mass, in the upper parts of it, settles down, freely and evenly, into the central parts of the kiln; whereas, in a conical furnace, the regular contraction of its width, in the upper as well as the lower parts of it, prevents the burning materials from settling uniformly, and levelling downward. They 'hang' upon the sides of the kiln, and either form a dome at the bottom of the burning mass, with a void space beneath it, thereby endangering the structure, if not the workmen employed; or, breaking down in the centre, form a funnel, down which the unburnt stones find their way to the draft holes. The contraction of the lower part of the kiln has not the same effect; for, after the fuel is exhausted, the adhesion ceases, the mass loosens, and, as the lime cools, the less room it requires. It therefore runs down freely to the draft-holes, notwithstanding the quick contraction of the bottom of the kiln or furnace.

Lastly, "That, with respect to the lime furnace, the fire requires to be furnished with a regular supply of air. When a kiln is first lighted, the draft-holes afford the required supply. But after the fire becomes stationary in the middle, or towards the upper part of the kiln (especially of a tall kiln), the space below is occupied by burnt lime, and the supply from ordinary draft-holes becomes insufficient. If the walls of the kiln have been carried up dry or without mortar, the air finds its way through them to the fire.

"In large deep kilns that are built with air-tight walls, it is common to form air-holes in their sides, especially in front, over the draft-holes. But these convey the air, in partial currents, to one side of the kiln only, whereas, that which is admitted at the draft-holes passes regularly upward to the centre, as well as to every side of the burning mass; and, moreover, tends to cool the burnt lime in its passage downward, thereby contributing to the ease and health of the workmen.

"Hence it is thought, that the size of the draft-holes ought to be proportionate to that of the kiln and the size of the stones taken jointly (air passing more freely among large than among small stones), and that the required supply of air should be wholly admitted at the draft-holes. By a sliding or a shifting valve, the supply might be regulated, and the degree of heat be increased or diminished, at pleasure."*

In opposition to the foregoing recommendations, it is held, that, although a (partial) reverberation is effected, the opening at the top of the kiln is sufficiently large to admit of a great escape of heat. But, admitting, that a sufficient reverberation takes place as to throw back the heat upon the limestone, it also throws back the *carbonic acid gas* which the main object of the calcination of the limestone is to expel; add to which, the uninflammable nature of the gas alluded to, serves to extinguish the flame.

If, in other respects, a trifling advantage may appear, we feel persuaded that the difficulty experienced by the workmen, in attempting to clear the *hollow* sides of an oval or elliptical kiln, will prove fatal to their general employment.† Irish kilns are filled with alternate layers of turf and limestone, of two feet and one foot in thickness, and are fired by igniting the turf at the bottom.

The only observation of general importance, made by lime-burners is, that the quantity of stone calcined, and the quantity of fuel used, depends upon

* Nicholson Arch. Dic. *Art.* "*kiln.*"

† A drawing of a kiln of the figure alluded to above, as given in Captain Smith's translation of "Vicat on Cements," and is there instanced as "*a bad form of kiln.*" See plate 1, fig. 9, of that work.

the *quality* of that fuel. Lime-kilns have, accordingly, been constructed with greater regard to the quality of the fuel employed, than to the *nature* of the stone to be calcined, or the operation of different descriptions of fuel upon the various species of limestones, during the period of their calcination.

Captain Smith observes, "That the effects of calcination are not confined to driving off the water of crystallization and carbonic acid from compound limestones ; it further modifies the constituent oxides, one by the other. In fact, if we treat an argillaceous carbonate of lime by a weak acid, it forms a deposit more or less abundant. After calcination, on the contrary, a complete solution is effected ; the clay, therefore, has entered into combination with the lime."

The important fact pointed out by Dr. Gilby, relative to the variety of character exhibited in the same mass of limestone rock, or in contiguity thereto, induces us to express a doubt of the possibility of constructing a lime-kiln capable of calcining the *whole mass of materials* in a perfect manner. It is, probably, owing to the unequal distribution of the constituent parts of the limestone in the same mass, that unburned pieces, dead lime, sub-carbonates, vitrified masses, &c., are formed ; for it is evident, that, to calcine the whole mass *perfectly*, the degrees of heat must be as varied as the quality of the materials. When the quality of the limestone is well ascertained, and found to be of uniform character (as in the case of Dorking lime) the experience derived from actual practice will soon point out the necessary temperature required for its proper calcination, and the kind of fuel best adapted for the purpose.

In order to point out the nature and peculiarities of some of the most important limestones indigenous to this country, we subjoin the following particulars, which have been deduced from actual experiment. The various specimens were procured on the spot, and each chosen as *ordinary examples of the kind* to which they refer.

OBSERVATIONS ON VARIOUS BRITISH LIMESTONES; ACTION OF ACIDS THEREON, &c.

Dorking Limestone—(Chalk).

This limestone consists of an indurated chalk marle (clay and chalk) of a dull grey colour, is opaque, and in minute, soft grains. When treated with muriatic acid previous to calcination, it effervesces violently and deposits a considerable quantity of clay (according to Smeaton, one-seventeenth part of its weight). After calcination it is light, friable, and of a yellowish tinge. If calcined sufficiently, and immersed in dilute muriatic acid, the *whole* is taken up, or held in solution.

This limestone is more or less intermixed with extraneous matter, depending upon the nature or peculiarity of the adjacent strata from which it is obtained. Thus, the chalk limestone at Greenhithe, Gravesend, and Dover, is associated with flint; that at Merstham with clay sand and a description of sandstone, while the Dorking limestone contains a proportion of clay.

Some ordinary specimens of Dorking lime, taken from Messrs. Bothwell's kilns, and slacked soon afterwards into a paste, formed a light cream-coloured tenacious mixture, slightly effervescent with acid. The lime evolved great heat in slacking, and rapidly fell to powder. When the Dorking limestone is reduced from the state of a carbonate to that of *lime*, by calcination, it should be used immediately, in consequence of its falling to powder by exposure; its strength is much impaired thereby, and more lime is required for its conversion into good mortar.

The peculiar fitness of Dorking lime for the ordinary purposes of building is noticed further on.

Kentish Rag—(Green Sand formation).

This stone is to be found abundantly in the Weald of Kent, Surrey, Sussex, &c. It is compact, hard, and of a dark gray colour. Effervesces violently in dilute muriatic acid,* and deposits a considerable portion of sand after the lime has been taken up by the acid. The greater part of the sand is large, and in transparent, colourless, and angular grains. It is interspersed, however, more or less, with green and red particles, of a silicious nature. After the effervescence the acid remains turbid.

This stone forms a pleasing ornamental masonry when properly displayed. It is durable, and forms excellent quoins for flint walls.

In the county of Kent, a cubic yard of this stone is called a *cord*. It requires twenty cords of stone to form a rod. The rod = 272½ feet super one foot thick. This masonry requires one hundred of lime and two yards of sand to the rod. Value per rod 13*l.* Worked quoins are of the value of 1s. 6d. per foot, labour only.

Portland—(Oolitic).

This stone is procured from the Isle of Portland. Similar varieties are also procured at Tisbury (Wiltshire), and Aylesbury (Bucks). The Portland stone effervesces violently in muriatic acid, and deposits a remarkably small portion of very sharp and fine-grained transparent sand. After the effer-

* In all these examinations the author has employed dilute muriatic acid as the solvent, and the products were examined with the microscope. Each specimen, before its dissolution, was about the size of a hazel nut.

vescence has ceased the acid is left tolerably clear. Portland stone chiefly consists of small egg-shaped particles of carbonate of lime closely connected together, and of a dull white or gray colour. After the Fire, in 1666, the quarries in the Isle of Portland were extensively worked, and supplied stone for the masonry of nearly all the buildings erected by Wren, and subsequent architects.

The employment of Portland stone for building has decreased materially within the last quarter of a century. It has been superseded by Bath stone, another of the oolitic series, which we shall presently notice. The action of the atmosphere upon Portland stone is very destructive, in consequence of the more soluble parts of the carbonate of lime becoming decomposed under its influence. The organic remains with which the stone abounds are consequently loosed from their matrix and fall out. The internal structure of the stone, becoming exposed, is now attacked, and rapid decay follows. In London, the effect of the atmosphere upon Portland stone is in other respects very remarkable. The projections of cornices, soffits, throatings, carvings, &c., are, in the interval of a few years, coated with a black stalactical substance, giving that dingy appearance so often remarked as being peculiar to the public buildings of the metropolis. Upon testing a quantity of this substance with muriatic acid a strong effervescence took place leaving a considerable residuum of carbonaceous matter. We offer the following as the most probable cause of this phenomenon:—

To a certain height, the atmosphere of London is strongly impregnated with carbonic acid gas, and also carbonaceous matter floating therein; the former being carried up in a heated state from the numerous fires, &c., combines with the water (particularly in the case of rain) of the atmosphere, and is thrown by the wind, &c., against the buildings. Now, it is known that water containing carbonic acid gas, becomes a solvent for *carbonate of lime*, consequently when the impregnated water settles, a slow decomposition will take place, and the loosened and undissolved particles of the stone, accompanied with the carbonaceous matter, will flow down the weatherings, &c., and settle upon the lower edges and surfaces of the ornamental or other projections, when it is dried, and forms the black mouldering substance above alluded to. To strengthen the foregoing supposition, we may remark, that in proportion to the height of the building, so are these effects more or less observable; the more lofty buildings, such as church towers, &c., being comparatively free from the black substance, near their summits, but increasing as we approach the level of the house tops. A few years ago, an attempt was made by Mr. Henning, the sculptor and medallist, to protect exposed specimens of sculptured masonry, by coating the stone with a peculiar composition (we think it was *wax and oil*). How he succeeded we are unable to say, but are certain, that an effective composition for the above purpose would prove of incalculable value in staying the rapid progress of decay, to which all buildings, constructed with limestone masonry, are particularly liable in this humid and changeable climate.

Bath Stone—(Oolitic).

The following varieties of Bath limestone are oolitic in their structure, and are carbonates of lime, having but little extraneous matter.

The whole of the series consist of minute egg-shaped grains of carbonate of lime, imbedded in a matrix of a softer nature, in which is mixed an immense number of very small comminuted shells.

Farley Down Stone.—Cream-coloured, effervesces violently in acid, and leaves a dark-coloured residuum. After effervescence the acid is left turbid, and of a dark yellow colour. It is unfit for damp situations, such as plinths.

Box Scalit, No. 1.—Lighter coloured than Farley Down Stone—effervesces strongly, and leaves little residuum. It contains small crystals of carbonate of lime. This stone is unfit for exposure.

Box Scalit, No. 2.—Cream-coloured, coarsely oolitic, strongly effervescent, and leaves little residuum. Unfit for exposure.

Box Corn Grit.—Cream-coloured, coarsely oolitic, intermixed with larger particles of shells, effervesces strongly, leaves no residuum. This stone is adapted for works *above ground.*

Box Ground Stone.—Coarsely oolitic, (some much flattened) dark cream-coloured, effervesces strongly, leaves more insoluble residuum than the others. This stone is a good weather stone.

Coombe Down Stone.—Coarsely oolitic, darker coloured than the above, with small shining crystals of carbonate of lime. This stone effervesces strongly, and leaves little residuum. It is a good weather stone.

In addition to the foregoing specimens, procured from the quarries, another specimen was examined, but with very different results. This specimen was obtained by the author, at Worcester, during the present summer (1838). It is a fragment from a block of Bath stone, that was being prepared for the restoration of the pinnacles at the eastern front of the Cathedral. The stone bears every resemblance to the *Box Corn Grit;* it effervesces strongly, but deposits a considerable number of oolitic particles, perfectly insoluble. These particles, when examined with the microscope, were observed to be of a yellow colour, semi-transparent, (in the moist state) and sufficiently tender to be crushed by a fine steel point. In the dry state the same particles were opaque and white. Another piece of the same stone gave the former residuum, with the addition of a few grains (apparently) of silex. The action of acid on the oolitic grains of the *carbonate* indicated that the grains consisted of two or more concentric *layers of lime*, as it could be distinctly perceived that each layer was, in its turn, attacked by the acid. Whether each of these grains have a nucleus, or, whether some are only hollow, we have not, as yet, been able to determine.

To ascertain the cause of the form and deposit of oolitic particles in limestone, is a subject well worth the serious attention of the practical chemist. The investigation of this subject might probably serve to account for the production of *boiler balls* in some steam-engine boilers.*

Brodsworth Limestone—(Oolitic).

Brodsworth Stone.—This limestone is procured from Brodsworth, in the county of York, 5½ miles from Doncaster. It is a tender limestone, of a light cream colour, works to a good face, and forms good ornamental masonry. In its internal structure, it approaches to the soft powdery appearance of chalk, but is slightly oolitic. It effervesces violently in acid, leaves little residuum, the acid is left turbid of a dark yellow colour, with a scum floating on the surface.

* For a description of these remarkable deposits, see "*Armstrong on Steam Boilers*," Part iii., pp. 77-8.

Purbeck—(Wealden Group).

Purbeck.—A very compact, smooth-grained, calcareo-silicious stone, effervesces in acid, and yields a considerable quantity of mixed sand, namely, rounded and angular. The sand is varied in colour, being composed of brown, green, and transparent grains, the latter particles being the most numerous. After effervescence the acid remains tolerably clear. This stone is practically so well known that its uses need not be pointed out.

Lias, from near Bath.

Colour, dark gray, effervesces strongly in acid, depositing a dark-coloured clay. The solution contains much iron, is very turbid, and of a dark green colour. A few hard black grains of a metallic lustre remain refractory in the liquid.

No. 2.—A very compact specimen of lias, from near Stoke Prior, Worcestershire, afforded, under a similar examination, corresponding results. Many specimens of the lias limestone form a very durable masonry. The *pedestal* of the great chimney shaft, at the British Alkali Works, at Stoke Prior, is built with this kind of stone. The pedestal is 75 feet in height, 36 feet diameter at the base or plinth, and 32 feet diameter above the plinth, which is ten feet six inches in height. The whole height of the shaft is 308 feet.

Septaria from the London Clay,*

May be properly classed among the hydraulic limestones indigenous to this country. It consists of "ovate or flattish masses of argillaceous limestone," arranged in nearly horizontal layers, chiefly found imbedded in the *London clay.* This substance is sometimes coated with a calcareous spar, or sulphate of barytes; it forms the basis of some well known hydraulic cements of good quality. A specimen procured from the excavation of the *Primrose-hill Tunnel* (London and Birmingham Railway), effervesced considerably in dilute acid, and deposited a greater quantity of clay than *lias;* the colour was also much darker.

Roche Abbey stone, (carboniferous series) is a compact crystalline limestone, of a gray colour. It effervesces slowly in acid, without leaving any residuum, and the solution is but little discoloured. This stone works to an excellent face, and forms durable masonry.†

Previous to the institution of certain experiments on *water cements,* by the late Mr. Smeaton,

* Also called Sheppy and Harwich Cement Stone, from its being obtained in considerable abundance at those places.

† A complete analysis of the principal British limestones will be given when treating of the quality of lime they afford.

an opinion had prevailed among practical men, that calcined chalk produced the weakest lim , in consequence of its being obtained from the most tender limestone.

To prove the truth or fallacy of this opinion, experiments were made with two specimens of limestone, apparently of very opposite qualities, the one being a piece of common chalk* and the other an equal quantity of *Plymouth limestone;* the result proved that no perceptible difference existed in the quality of the two limes, but that each was of the same crumbling and weak nature; satisfactorily proving, that the quality of the lime was not owing to the *density* of the limestone. Succeeding experiments, made by the same engineer, established a fact of much greater importance, namely, that *pure limestones*, or those carbonates of lime that dissolve in acids without residue, are almost useless for the *manufacture of building lime*, when compared with those that are alloyed with a certain portion of *clay*. On the subject of *Dorking lime*, which has an alloy of this kind, he thus speaks, " From the experiments now related, I was convinced that the most *pure* limestone was not the best for making mortar, especially for building in *water:* and this brought to mind a maxim I have learnt from workmen, that the best lime for the *land* was seldom the best for

* *Common Chalk* is nearly pure. *Chalk Marle* is a very different substance, although belonging to the same geological formation.

building purposes; of which the reason now appeared, which was, that the most pure lime affording the greatest quantity of *lime salts*, or impregnation, would best answer the purposes of *agriculture*: whereas, for some reason or the other, when a limestone is intimately mixed with a proportion *clay*, which, by burning, is converted into *brick*, it is made to act more strongly as a *cement*.* This suggested to me the idea, that an admixture of *clay* in the composition of limestone, when treated as above specified, might be the most certain index of the validity of a limestone, for *aquatic buildings*; nor has any experience since contradicted it, as all the limestones in repute for water works, that I have met with, have afforded this mark. Even the *Dorking* lime, much esteemed for those uses in *London*, and in the country round about, is plainly nothing but a species of chalk, impregnated with clay, of which it makes full one-seventeenth part of its original weight."

The ordinary class of calcareous minerals furnish the following descriptions of lime upon calcination:—1, *rich* lime—2, *poor* lime—3, *slightly hy-*

* It must not be wondered at that workmen generally prefer the more *pure* limes for building in the *air*, because, being unmixed with any uncalcareous matter, they fall into the finest powder, and make the finest paste, which will of course receive the greatest quantity of *sand*, (generally the cheaper material) into its composition, without losing its toughness beyond a certain degree, and requires the least *labour* to bring it to to the desired consistence: hence, mortar made of such lime is the least expensive; and in *dry work* the difference of hardness, compared with others is less apparent."—*Smeaton's Narrative, &c. of Eddystone Lighthouse.* Fol. Edit. 1793. p. 108.

draulic lime—4, *hydraulic* lime—and 5, *eminently hydraulic* lime. The quality of these limes vary with the proportion of *silex, alumen,* iron, &c., &c., contained in the stone; and the quality of the lime is further modified by the degree of calcination to which the specimens are subjected.* Our present observations are confined to two of the above kinds, namely, the *hydraulic* lime, and the *rich* lime.

The annexed series of British limestones† are, from the proportion of the alloy in their composition, eminently calculated to produce strong hydraulic lime upon calcination. It will be observed that they are chiefly of the lias species.

No.	Species of Limestone.	Proportion of Clay.	Colour of the Clay.	Reduction of weight by burning.	Colour of the Brick.
1	Aberthaw	$\frac{3}{23}$	Lead colour.	4 to 3	Gray stock brick.
2	Watchet............	$\frac{3}{25}$	Ditto	4 to 3	Light colour, reddish hue.
3	Barrow	$\frac{3}{14}$	Ditto	3 to 2	Gray stock brick.
4	Long Bennington ..	$\frac{3}{22}$	Ditto	—	Dirty blue.
5	Sussex Clunch	$\frac{3}{16}$	Ash	3 to 2	Ash colour.
6	Dorking............	$\frac{1}{17}$	Ditto		
7	Berryton Gray Lime	$\frac{1}{12}$	Ditto		
8	Guildford	$\frac{2}{19}$	Ditto		
9	Sutton (Lancashire)	$\frac{3}{16}$	Brown		

* See also "Vicat," on Mortars and Cements. Translated by Captain Smith, p. 9.

† Extracted from Smeaton's Narrative, &c.

Effect of Iron on Calcareous and Silicious Matter.

Iron appears to be a very active and energetic agent in the formation of natural cements. Its combination with some of the calcareous, and nearly all the silicious deposits, exhibits much diversity of character, and, in many instances, can only be separated by careful analysis aided by considerable chemical experience. The lias deposit contains iron in considerable quantity, but the calcareous rocks above the lias (with the exception of the lower Bath oolite) contain little oxide or carbonate of iron.* In *sub-aqueous* works, it is highly necessary to *increase* the hydraulic properties of the lime employed, by the admixture of certain substances whose constituents will operate energetically upon it. These substances, whether natural or artificial, are termed *Puzzolanas*.* The expense attending the importation of the natural *volcanic Puzzolana*, from Italy, almost prohibits its employment in the engineering works of this country, but several efficient substitutes

* See "Phillips's Treatise on Geology," vol. i. p. 202.

† This mineral is found in the greatest abundance at Puteoli, in Italy (now called *Puzzuoli*). It is of the *lava* genus, magnetic, and easily melts *per se* into a black slag. It suddenly hardens when mixed with one-third of its weight of lime and water, forming a cement more durable under water than any other. Bergman found one hundred parts of it to contain fifty-five to sixty of silicious earth, twenty of argillaceous, five or six of calcareous, and from fifteen to twenty of iron: this last constituent is considered to be the cause of its property of hardening under water. The iron decomposes the water of the mortar; and thus, in a very short time, a new compound is formed.—*Stuart Arch. Dic.*

have been discovered, in all of which, *iron* forms a material constituent; the composition, mode of preparation, &c. of these *artificial Puzzolanas*, will be subsequently adverted to. The admixture of pulverized *forge scales*, or *iron dross*, with calcareous matter, produces a very strong hydraulic mortar. A composition of this kind was used at the works of the Eddystone Lighthouse, and appears to have originated with the Engineer (Mr. Smeaton) from observations incidentally made on the cementing property of iron when disseminated in a gravelly or silicious matrix.

Having recently witnessed an extraordinary instance of the effects of iron when in contact with silicious matter, we now direct the attention of our reader to the following particulars: The foundations of the piers of Westminster Bridge were originally laid in *caissons* in the following manner:— A massive raft of timber was constructed of a form and size suitable to the pier; furnished with a bottom, ends, and sides, but the ends and sides were secured together in such manner that they might be readily detached from each other when necessary: a portion of the pier was erected in this sort of flat bottomed barge; and, when all was in readiness, water was admitted in order to sink it, the descent being regulated by guide ropes: the load of masonry having been deposited on the bed of the river, (previously levelled and prepared to receive it), the *sides* and *ends* of the *caisson* were withdrawn from the raft, to be again used as occasion might require: the piers were protected externally, around the bases, by permanent piling, driven into the bed of the river, their lower extremities being pointed in the usual manner and shod with iron.* The defective nature of a foundation thus constructed became evident previous to the completion of the structure; since that period repairs have been frequent and extensive, and at the present time some considerable operations of this kind are being effected by Messrs. Cubitt. In the course of the works it became necessary to withdraw some of the fender, or sheeting piles, which had been driven into a gravelly substratum, and it was found that their lower extremities were coated with a concreted mass of coarse flinty gravel, of extreme density. Upon examining this concretion, we found that the greater portion of the *iron* with which the feet of the piles were protected had oxidised and combined with the gravel. On fracture, the substance exhibited the same appearance as iron-sand, or stone, when strongly impregnated with the metallic oxide; when treated with dilute muriatic acid it effervesced considerably, and evolved an odour similar to *sulphuretted hydrogen;* lastly, a pulverized portion of the concrete having been thrown into *lime water*, restored the solution *(lime water)* to a pure state. From the foregoing examination, we deduce the following conclusions:—

* See Mechanic's Magazine, vol. xiii. p. 322.

1st. That iron, imbedded in a gravelly and silicious stratum, will oxidise and dissemininate its decomposed particles in the surrounding matrix, forming a ferrugino-silicious concrete of extreme density.

2nd. That a concrete thus formed becomes an energetic *artificial Puzzolana,* indicated by its effervescing with acid and decomposing lime water.

It is highly probable that the decay of the foundations at Westminster Bridge may have been, in some degree, retarded by the formation of this concrete around the base of the piers, and the pointed ends of the piles.

On the Use of Rich Lime.

Those limestones which dissolve freely without discolouring the acid, and leave little or no precipitate after the effervescence is completed, are generally capable of furnishing *rich lime.* Calcined shells produce a lime of this description, and we are informed by the author of the " Parentalia," that the mortar used for the " *under drawing and inside work* " of " St. Paul's Cathedral " was composed of " *calcined shells,*" and produced excellent mortar, This statement induced the late Mr. Smeaton to investigate the quality of " *shell lime* ;" upon trial, he found it to set hard, and readily, without any admixture of sand, tarras, or other matter. In short, for water work, tarras scarcely appeared to improve its natural quality. On being put into water, after it was set, it did not dissolve, but did not acquire an additional hardness; on the contrary, by degrees it macerated and dissolved, not internally, but gradually from the surface inwards; and hence he concluded it totally unfit for use. He was afterwards informed that a part of the work at *Ramsgate Pier* had been done with this kind of lime, but was subsequently taken up, on its dissolving quality in sea water being discovered.* A recent examination of the principal buildings erected by Sir C. Wren has satisfied the author that, although *rich lime* was extensively employed in their construction, there yet remains sufficient evidence to show that the architect did occasionally introduce materials, in conjunction with the mortar, of somewhat remarkable character.

During the recent repairs of the parish church of St. Vedast, Foster-lane, an opportunity was afforded to institute a minute examination and admeasurement† of the tower and spire, completed by Sir C. Wren, in the year 1697. In the course of the examination, particular attention was directed to the state and quality of the mortar originally employed in that building. The vertical joints of the masonry composing the *spire* were considerably

* Smeaton's Narrative, &c., p. 106.

† For the kindness of the architect, Mr. S. Angell, in permitting the necessary survey to be made, the author begs to return his sincere acknowledgments.

opened (probably caused by the expansion of the metal cramps), and the mortar was found to be in a weak and crumbling state, but the base of the *obelisk* terminating the spire, was discovered to be laid on a thick mortar joint in which was imbedded two courses of *flat oyster shells,* these together with the mortar, had remained uninjured for upwards of *two centuries and a half.*

We are not yet prepared to offer a satisfactory solution of this phenomenon, the necessary experiments will require considerable attention and occupy much time. For the present, we refer the reader to the note* at the bottom of the page, in which is contained some observations that may possibly supply data for a further investigation.

* "It is remarkable that oysters, and shells which, like them, are composed of distinct broad lamella of alternating membrane and carbonate of lime, have resisted in almost all rocks, argillaceous, calcareous, arenaceous, the chemical changes to which venerida, trigoniæ, and others of an apparently compact texture, have completely yielded. While the former retain their lamella and pearly surfaces, the latter have often been wholly dissolved in limestone rocks, and their places left vacant; while a cast of the inside of the shell, and an impression of the outside, disclose completely the history of the change. A further process is frequently superadded, by which the cavity is again partially or wholly filled with crystals of carbonate of lime, which has been introduced by filtration through the surrounding rock. In other cases, silicious matter, pyrites, and other substances, have passed by a similar process. The common fossil, called belemnites, of the same group as the cuttle, is a remarkable instance of the force of the original structure in controlling the effects of chemical agencies; for in clay, sands, chalk, flint, limestone, pyrites, this singular fossil generally retains its fibrous structure, colour, translucency, and chemical properties; while in the same masses echini are changed to calcareous spar, and sponges to flint, and many shells have totally vanished."—*Phillip's Geology*, vol. i., pp. 78-9.

The remarks made by Mr. Smeaton, on the quality of rich lime, have been confirmed by the more recent experiments of M. Vicat, and it is now decided that *rich lime* is incapable of furnishing a good mortar, and that the admixture of quartzose, silicious sands or other *inert* substances so beneficial to hydraulic limes, is positively injurious to *rich lime.**

* Captain Smith states, that the quality of *rich lime* is much improved by adding a small quantity of *coarse sugar* to the water with which it is to be worked up.—(See his Translation of Vicat, note, pp. 84-5.)

CHAPTER VI.

ON THE PROPORTION AND ADMIXTURE OF THE INGREDIENTS EMPLOYED IN THE COMPOSITION OF ORDINARY MORTAR. QUALITY OF SANDS, SAND-STONES, &c.

AMONG the recent investigations which have materially assisted in promoting a more perfect theory of the consolidation of mortars, those of Professor Fuchs are entitled to hold a prominent place. The various compounds of lime and sand, constituting the materials of common mortar, and the admixture of unslacked lime, with broken stones, ballast or gravel, forming the basis of the strongest concrete, severally possess certain properties, and exhibit distinct phenomena, referrible to a most important chemical and mechanical combination.

In the course of his investigations, Professor Fuchs noted certain facts, from which he concluded, that the induration of various kinds of mortar depended upon the formation of silicates of lime, and sometimes also of alumina of silicates. He discovered that these silicates retain the water and acquire the hardness of masses of stone, while the hydrate of lime in excess is gradually united with carbonic acid, so that the indurated mortar may be considered as a compound of carbonate of lime and zeolite.

Opal, pumice-stone, obsidian, and pitch-stone, simply pulverised, form a good cement with hydrate of lime, while quartz and sand only produce an hydrated silicate upon the surface of each grain. Although the mass is thus connected, it does not readily become solid. If the quartz has been reduced to a fine powder, the more solid will the mass become. Now, if one-fourth part of lime be mixed with the quartz and the whole be well calcined, so that the mass becomes a frith, and the whole be afterwards pulverized and mixed with one-fifth part of lime, an hydraulic mortar or cement is obtained, which attains a sufficient hardness to admit of being polished. Felspar with lime hardens slowly, and only at the end of five months, but if calcined with a small quantity of lime it becomes much better. Water abstracts from this mortar six per cent. of potash. Common potter's clay, which is worth absolutely little or nothing when uncalcined, produces, when calcined with lime, a cement which hardens perfectly well, provided it contains only a small portion of iron.

Professor Fuchs also found that steatite,* which had been heated to a bright red heat, would not combine with lime, and, concluding that magnesia possessed a very strong affinity for silicious acid, he

* Soap stone, a sub-species of rhomboidal mica—*Ure*.

tried the employment of calcined dolomite (magnesian limestone) for the cement instead of common lime, and found that it greatly surpassed the latter, both for the preparation of common mortar and for that of hydraulic mortar. He also obtained good mortar of the latter kind with calcined marl.*

The greater number of specifications prepared by surveyors for the erection of buildings, direct that the mortar shall be composed of stone-lime and sharp river-sand, to be mixed in the proportions of one part of lime to three parts of sand. These proportions will make excellent mortar if properly compounded; but, as the quality of the lime varies considerably, so will it take more or less sand.

Builders employ two methods of compounding their mortar:—First, when it is required to convey it in a dry state to the work, it is done by forming a bed of lime, surrounding it with sand, and then throwing on the lime a sufficient quantity of water to slack it, and covering it up immediately with sand; after it has remained some time in this state, it is turned over, and, if necessary, screened. The mixture is now in the state of a dry powder, and can be carted to the work, where more water is added and it is chafed up for use. The other method is employed when there is convenience for making it up at the work. In this case it is what is termed "*larryed.*" Thus:—the lime is put into the middle of a bed of sand, and a large quantity

* See Repertory, 1837.

of water thrown on, and with lime-hoes mixed up immediately until completely incorporated. It is then allowed to remain for a few hours, when it becomes set, and of proper consistency for use. The lime when turned up in this way will admit of a larger quantity of sand, as all the particles of lime are dissolved, whereas by the first method, there are always small particles of the lime that cannot be properly mixed, however much it may be chafed up. From observations recently made, it was found that seventy-two bushels of stone-lime required eighteen yards of sand, making good "*larryed mortar*," the cubic contents of which = 315 feet.

Chalk-lime mortar requires two parts of lime to three of sand, and is now chiefly used for plasterer's work.

The theoretical investigations connected with the composition of mortars, have for a considerable period occupied the attention of the most celebrated French chemists. In researches of this nature they are considerably in advance of us; but the reputation of the English builders for *practical experience* in the manipulation of mortars is, on the other hand, well known and appreciated by all those whose attention has been directed to these important investigations. The professional practice of English architects, relating to the composition of mortars employed in public or other works, is best exemplified by extracts from their original specifications. We subjoin a few, which will show the relative degree of importance attached by different architects and engineers to this division of their instructions, and the proportions of materials, mode of manipulation, &c., adopted by each.

Ordnance Works.

Not less than thirty-nine bushels of lime* to every rod of brick-work. The lime to be slacked under cover, and the mortar made on a stone or brick-floor, and properly worked until all the parts are thoroughly incorporated.

Farm-house and Farmery. Erected at Greenford, 1832.

Good fresh gray lime and river-sand properly screened, and in the proportion of *one of lime to three of sand,* well tempered together.

Stockport Church, Cheshire.

Brixton lime (Cheshire) *mixed with two-thirds of sand.*

* Thirty-nine bushels of lime = one and a half cwt. of lime, and would require seventy-five bushels of sand to make enough mortar for a rod. In the printed contract no distinction is made as to chalk or stone-lime.

Bankrupts' Courts, Basinghall Street, London.

The mortar to consist of *one-fourth of good Dorking lime, and the remaining three-fourths, of clean sharp river-sand.*

Turkey Street Bridge, Enfield, 1827.

The brick-work to be laid in Dorking stone, or blue lias lime-mortar, and clean sharp Thames sand, in the proportions of *three parts of sand to two of lime;* to be made in a pug-mill and used hot. The grout to be in the same proportions.

Exeter Higher Market, Devon. 1836.

Mortar for the brick work is to be made with well burnt stone lime (Plymouth) and good clean sharp grit sand, in proportions of *one and a quarter to three.* The lime is to be thoroughly slacked with an abundant quantity of water as soon as it is brought on the premises, and immediately after, well protected from the action of the atmosphere by a thick covering of the ingredient with which it is to be mixed; in this state it is to lie until quite *cool,* and then mixed together in the above proportions. This mixture is afterwards to be sifted through a wire sieve and made up with such a quantity of water only as will render the compound of the consistency of damp sand. It is to be well tempered in a pug-mill, and to be laid in heaps from two to three weeks, being protected during this period so as effectually to prevent it becoming dry or setting. Ultimately it is to be worked up for use. The mortar for the free-stone is to be made up of fine lime putty, and bright coloured fine sharp sand, washed clean, and well tempered together.

Silk Mill and Engine House for Messrs. Grout, Baylis & Co., at Great Yarmouth, 1825.

The mortar for the mill, engine, and boiler house to be composed of the best Dorking lime and sharp sand, in the proportion of *three parts of sand to one part of lime.* The mortar for the chimney (one hundred feet in height) to be the same as before described, but for the other parts, to consist of *one-third part of the best chalk lime (Yarmouth) used fresh from the kiln, and two-thirds of clean sharp sand, well worked together.*

House in Orton's Buildings, Southwark Bridge Road, in the Liberty of the Clink, London.

Mortar to be composed of Merstham stone-lime and sharp drift sand from above London Bridge, in the proportion of *one heaped bushel of lime to two striked bushels of sand,* well tempered.

Works at the Berkeley and Gloucester Canal.

The mortar must be made of clean sharp sand and Aberthaw lime (Wales), in the following proportions: that is to say, for works under water, and exposed to the river Severn in front, *two measures of sand to one of lime;* for backing all work under water, and exposed to water, *three measures of good sand to one of lime;* and for all other backing, *four measures of sand to one of lime.*

The mortar required for the locks and the outer gates of the basin to be intimately mixed in a pug-mill. The limestone is to be brought to the ground and burned upon the site of the works, the contractor having the use of any kilns now thereon; but he is to erect others if necessary.*

Greenwich Railway.

The mortar to be composed of the best Halling lime and sharp river sand from above London Bridge, in the proportions of *one of lime to two and a half of sand.*

London and Birmingham Railway.

Mortar for all the works near the London station to be composed of the best burnt Dorking or

* Blunt's Civil Engineer.—Division A.

other lime of equal quality, and clean sharp river sand, in the proportion of *three of sand to one of lime;* the compost is to be mixed in a dry state, and passed through a pug-mill with a proper quantity of water.

Watford Tunnel.

The mortar to be used in the tunnel shall be made with the best fresh burnt Merstham or Dorking lime, or other lime which the Engineer may deem equally good. It shall be ground under edge stones in its dry or unslacked state. The sand must be sharp and clean, and mixed with the lime in the proportion of *three measures of sand to one of lime.*

Blisworth Division.

The mortar to be used in the beds and faces of buttresses, walls, sides, or drains, and invert arch, to consist of *one part of lime to three parts of clean river or other unexceptionable sand.* The sand to be passed through a quarter inch screen; the lime to be fresh and well intermixed by a thorough beating. The mortar for running into the upright joints of the courses, and for filling in the work sound, to consist of *one part lime to four parts of small unscreened gravel,* to be well mixed and beaten to a tough consistency, and liquified in tubs or other vessels, to be properly adapted to run into and fill up all vacuities.

The mortar to be used as hot as is consistent with the safety of the work, and the sand and gravel to be perfectly free from any loamy or other particles of a muddy nature.

The limestone rock, found in the excavation, may be used for the mortar specified to be used in the retaining walls of this contract.

The foregoing extracts from the original specifications, respecting the composition of mortar, exhibit a tolerable agreement with each other, and with the exception, that the preliminary directions of some, are more copious and explanatory than others, may be considered as affording a concise and satisfactory view of the modern practice adopted by some of the most eminent builders and architects, grounded upon careful observation, and great practical experience. The quality of the sand required for the composition of mortar, should undergo careful examination, as it operates with considerable influence on the quality of the mortar.

In districts where sand of the requisite energy and sharpness is unattainable, road-drift has been employed, and, in some instances, recommended in the specification of works. Where the repair of roads has been effected with a mixture of good flint and quartz gravel, and the drift thoroughly screened to separate the muddy and other extraneous particles, such sand may afford a tolerable substitute. But in cases where broken granite has been employed for road-making, the drift arising therefrom should in no wise be employed for the compounding of mortar.

Granite consists of distinct aggregations of *quartz*, *felspar*, *mica*, and *hornblende*, each in a crystalline form. *Felspar* is of a whitish, sometimes of a reddish, colour, quite opaque, and occasionally crystallized in a rhomboidal form; *quartz* is less abundant, somewhat transparent, and of a glassy appearance; *mica* is dispersed throughout in small glistening plates, the colour is dark and the appearance metallic;* *hornblende* (a simple mineral) is sometimes present, and is the same mineral which imparts the deep green colour to the Plutonic rocks, *greenstone and basalt.* The *road*-sand, derived from the pulverization of granite, is, generally speaking, a very unfit ingredient, and ought not to be employed in the fabrication of mortars. The trituration of the micaceous particles forms a dust of no coherence, and the harder particles become *rounded* by attrition, thereby destroying a form so essential to a proper adjustment when mixed with lime, viz. *sharpness* and *angularity*.

There is also a description of granite much used in the neighbourhood of London, for mending the roads, which also forms a road-sand of very inferior quality. This granite is of a very dark colour, almost black; it essentially consists of *mica*

* Jamieson, "Mechanics of Fluids," note G, page 459.

and *white felspar*, the former in great abundance. When pulverized, it becomes a gray powder; tenacious, and clay-like, while wet, but drying to a fine micaceous dust, containing a portion of iron.

In the neighbourhood of Great Malvern, the roads are repaired with the *sienitic granite* of the locality. The sand, produced from the decomposition of a portion of the rock, is red,* opaque, and angular, in large grains. When washed it yields a small portion of reddish-coloured residuum, which contains iron. Acid has no effect upon the stone.

The sand mostly used in London and its vicinity for the manufacture of good mortar, is procured from the bed of the river Thames, above the bridges. This sand has acquired a standard reputation among the principal architects and builders of London, and we shall presently show, that its good qualities have not been over estimated by the profession. The river Thames passes through a country where the contiguous strata is calcareo-silicious, and argillaceous. The immense number of weirs, culverts, road-drains, &c., which are discharged into the river, deposit an immense mass of heterogeneous matter, consisting of calcareous, *fossil*, *quartzose*,† and *flint-sands*, particles of coal alluvium, and much *iron*.‡ Sea-sand, flints, and marine debris is also washed backwards and for-

* A very considerable portion of the rock is composed of red felspar.

† From the pulverized particles of road gravel, which contains much quartz.

‡ Iron mixed with the road-sand. In Mr. Babbage's interesting work on the "Economy of Machinery," it is stated, that "every coach which travels from London to Birmingham distributes about eleven pounds of wrought iron

wards with the flux and reflux of the tide, dis tributing a sand of coarser quality than that deposited by the inland tributaries. The Thames sand, therefore, requires to be well screened and washed, previous to its admixture with the lime, in which state it will be found to consist of *two* kinds of materials—the first a fine *shingle*, mixed with calcareous particles, and the second a fine angular, *silicious* and *quartzose* sand. The accidental circumstances which occasion the deposit of a fine and coarse sand in the bed of the river Thames are the chief causes of its superiority as an ingredient in the composition of mortar, for, the interstices between the larger and more quartzose particles, become partly occupied by the angular and sharp fragments of the finer sand, which firmly unite with the cement, wedging and dove-tailing them together.

It is almost unnecessary to observe, that *washed sea-sand* will produce precisely the same effects.

Thames ballast, (used for concrete), is an admixture of sand and small stones, charged with the impurities mentioned above; it is procured from below Blackfriars Bridge. A sample of this mixture, tested with acid, effervesces considerably, and deposits much clay containing iron. A washed fossil, or pit-sand, of very fine quality (gray colour)

along the line of road between those two places." The result of this calculation was derived from observations made by Mr. Macneill, the superintendant, (under the late Mr. Telford) of the Holyhead roads.

used by plasterers, also effervesces with acid. Silver sand, for *fine stuff*, is procured from the Isle of Wight. This sand is angular, transparent, and colourless; acid has no effect upon it, and it is perfectly free from impurities.

The fossil and land-sands of Bagshot Heath, Highgate, and Hampstead, (when washed and cleansed from the impurities which they contain) are employed for internal work, such as plastering, &c. These sands are exceedingly numerous and diversified; the quantity of extraneous matter contained in some samples, is shown by the following

*Analysis of Three Samples of Bagshot Sand.**

No. 1.

Silicious Sand	82
Alumine	2
Oxide of Iron	7
Lime	0
Vegetable Matter	9
	100

No. 2.—(*Partaking of a Clay appearance.*)

Silex and Silicious Sand	85
Alumine	6
Oxide of Iron	5
Lime	1
Vegetable Matter	3
	100

* From some Specimens in the "Polytechnic Institution," Regent Street.

No. 3.

Silicious Sand	90
Alumine	1
Oxide of Iron	4
Lime	2
Vegetable Matter	3
	100

A section afforded by the excavations made during the progress of the works at the North London Cemetery, Highgate, induced us to examine some other specimens of these sands in detail. Immediately under the alluvium, is a coarse, gravelly, or craggy bed, in which is contained a broken stratum of indurated sand, highly impregnated with metallic oxide. The strata next in succession consists of a series of red sandy beds, much intermixed with loam, followed by a thick stratum of very fine light-coloured sand, nearly free from impurities. A cursory analysis of these sands, enables us to give the following particulars concerning their quality, &c.

The ferruginous mass of sand or *sand-rust*, (as it is provincially termed) is stratified and highly indurated; upon examination with a strong lens, the particles appear to consist of crystallized quartz, which in themselves are chiefly transparent and colourless, but are much stained by contact with the ferrugino-silicious cement which connects the mass. The external appearance of the sand is highly metallic, and of a dark brown colour; when pulverized and washed, it leaves scarcely any deposit, and is not affected by the action of acid. After maceration in dilute sulphuric acid, and the liquor being filtered, the iron is separated from the sand in the form of a dense blue precipitate, by adding prussiate of potass; or, a cloudy precipitate (iron) of a reddish brown colour, in considerable quantity, may be obtained, by adding a solution of pure ammonia to the impregnated acid. The red sand from the loamy beds, when washed and separated from the clay, consists of very fine seed-like transparent grains, of a reddish brown colour. The clay, or loam, is of a bright yellow colour, and very tenacious; it contains less iron than the previous specimen. The last specimen is a light-coloured glassy sand, nearly pure, and consists of finely comminuted angular fragments of silicious matter; it does not appear to contain iron, but is nevertheless admirably adapted for the manufacture of mortar.

A specimen of concreted sand procured from *Sandy, in Bedfordshire*, when microscopically examined, exhibited appearances analogous to the specimen of *iron-sand* before described. Considerable difficulty was experienced in cleansing this and the former specimen, from the metalliferous cement which connected the quartzose particles together. Frequent washing had no other effect than that of detaching the more earthy particles from the mass; but, upon macerating the sand in dilute sulphuric acid* for four or five days, the iron became separated, and the sand, after being washed, was left in nearly a pure state. The sand consists of large quartzose particles, colourless, transparent, and of very irregular figure, but much worn and rounded by attrition. The test of *ammonia* applied to the filtered acid, gave a more dense and copious deposit than any other specimen submitted to examination. The geological deposit to which this description of sand belongs,† contains a sand-stone of great hardness and durability. It is much used in Kent, Sussex, and some parts of Norfolk, as a paving and building stone. The groin work of *Battle Abbey*, in Sussex, is mentioned by Mr. Conybeare, as being constructed with a free-stone belonging to this series.

The *Sands*, derived immediately from the disaggregation of *Sandstone rocks*,‡ next demand our attention.

The following series of sand-stones are principally aggregations of quartzose particles, cemented together with a portion of clay; the clay consisting of silex, alumina, and iron, in combination.

* A more detailed account of the action of acids on clays, sands, &c. will be given in another part of this work. It will be perceived that our present observations chiefly apply to the quality of a few varieties of the quartzose sands and sandstones.

† Iron sand.

‡ The nature and quality of *sandstones*, and the uses to which they may be applied in engineering operations, will form the subject of a distinct chapter.

In one instance, only, was effervescence perceived, viz. in the specimen of stone, No. 1, from the Forest of Dean.

The stones were examined in the following manner :—

A specimen of each was pulverized and a portion of the sand was carefully washed in order to separate the aluminous powder, &c. Dilute sulphuric acid was now added and the sand was again washed and examined with a good microscope, to detect its quality. The clay being subjected to the action of the acid, in order to separate the iron, was left to macerate, for some time ; the acid from the clay was then filtered through white filtering paper, digested, and finally tested with a solution of pure ammonia, which precipitated the iron, &c. The clays were then dried and reserved for further examination.

Bramley Fall Sandstone, No. 1—(Yorkshire).

A brown-coloured sandstone of coarse grain. This stone is difficult to work to a good face, but is nevertheless a good weather stone. It is now much used in situations where granite was formerly employed,* and is in some respects superior to it, which will be hereafter explained. This stone pulverizes to a coarse brown sand. When separated

* Such as bases to columns and plinths, springing stones to arches, &c. &c.

from the clay, &c. by washing, it exhibits a sand composed of large rounded grains of quartz. Some of the grains are colourless and transparent, many are tinged with brown, and others are white and opaque. Acid has no effect upon the sand from this stone, but the clay, (remarkably small in quantity, and of a fine buff colour) yields a portion of iron when tested as above described.

Bramley Fall, No. 2.

A light-coloured specimen, finer grain than the above. The sand, when washed, appeared angular, transparent, and colourless. Clay, buff-colour, small in quantity, and yielded a slight trace of iron. Acid produces no effect upon the stone.

Holton Sandstone—(Yorkshire)

A light brown sandstone of fine grain. Stratification widely separated. The sand, when separated from the clay by washing, appeared in the form of fine, round, seed-like transparent grains, of a light brown colour, interspersed with micaceous particles. Acid has no effect upon this stone, but the clay (light drab colour) yields iron, in tolerable quantity.

Kirby Sandstone—(Yorkshire).

A drab-coloured sandstone of coarse grain. Stratification widely separated. When pulverized the sand is of a light brown colour. Washed and separated from the clay, the particles of sand are found to consist of angular quartzose grains, transparent and opaque, intermixed with scales of mica. The colour of the clay is light brown, and yields iron. Acid has no effect upon the stone.

Leeds Flagging—(Yorkshire).

Light drab-coloured sandstone of fine grain, closely stratified, but splits readily at the matrices or divisions, which are very smooth, and coated with mica. When pulverized, gives a fine light brown sand. Washed and separated from the clay, the particles of sand exhibit the appearance of very fine quartz-like grains, having a brown tinge, and particles of mica intermixed with the sand, colour of clay, dark yellow. The test (ammonia) gave little or no indication of iron. Acid has no effect upon the stone.

Mexborough Sandstone—(Yorkshire).

A drab-coloured sandstone of coarse grain, Stratification widely separated. When pulverized.

gives a fine light brown sand. Washed and separated from the clay, the grains appear like small fragments of gum, some of which are transparent. Colour of clay, yellow, and contains iron. Acid has no effect upon the stone.

Green Moorstone (light coloured)—(Yorkshire).

A dark drab-coloured stone, of very compact texture. Stratification indistinct. Exhales earthy odour when breathed upon, and sulphurous smell, when freely rubbed with an iron point. When pulverized gives a fine light drab-coloured sand, which forms a stiff tenacious paste, when mixed with a small quantity of water. The sand, when washed and separated from the clay, exhibits minute seed-like transparent grains, of a brown colour. When the acid from the clay is tested with a solution of ammonia, much iron is precipitated. The clay is of a drab colour. Acid has no effect upon the stone.

Moorstone (light coloured) *No.* 2—(Yorkshire).

A very compact stone, fine grain, gray colour, and indistinct stratification. When pulverized, gives a fine light gray sand. The grains of sand, when washed and separated from the clay, are of a dull white colour, transparent, minute, and rounded. Colour of the clay, dark drab, contains much iron. Acid has no effect upon the stone.

Idle Flagging—(Yorkshire).

A dark cream-coloured sandstone, closely stratified, and possessing a fine grain. When pulverized, gives a very light brown sand, or powder. Washed and separated from the clay, the grains appear in small, sharp, transparent, silicious particles, slightly tinged with brown. Colour of the clay, light brown, contains iron, which is precipitated slowly. Acid has no effect upon the stone.

Penshurst Sandstone—(Kent).

A tender sandstone, fine grain, deeply tinged with iron oxide throughout its substance. When washed and separated from the clay, the sand appears in fine, seed-like, red grains. Acid has no action on the stone. The colour of the clay is a dark brown, approaching to red, and contains much iron.

Sheffield-stone—(Yorkshire).

A light gray sandstone, fine grain, and compact texture. When pulverized, gives a light gray sand. Washed and separated from the clay, the grains of sand appear rounded, seed-like, and nearly colourless. Colour of clay nearly white, contains much silex and little iron. Acid has no effect upon the stone.

Sandstone, from the Forest of Dean.—No. 1.

A remarkably hard light gray stone, with a slight tinge of red. The grain is coarse, but the stone is capable of being worked to a smooth face. When pulverized, it gives a coarse pink coloured sand. Washed and separated from the clay, the sand appears slightly granitic, of various colours, viz., red, brown, opaque, colourless, and transparent. The colour of the clay is a dull red, and contains much iron. This stone effervesces slowly in acid.

Sandstone, from the Forest of Dean.— No. 2.

A very hard, light gray coloured sandstone, finer grain than the above, and works to a good face. When pulverized, gives a light drab coloured sand. Washed and separated from the clay, the sand appears in small quartzose grains, colourless and transparent. Colour of the clay, drab. Acid has no effect upon the stone.

* Lighter colour than Kirby-stone.

Each of the above specimens of sandstone are described as containing *clay*;* this ingredient, however, varies considerably in quality. In some of the specimens, the *alumina* is so small in quantity as to be with difficulty separated from the mass.

The relative value of the various sands, as ingredients in the composition of mortars, stands next in order for our attention.

* We had some doubt of the propriety of naming these deposits from the sand of sandstones "*clays*," as it was found, that, when perfectly dry, most of them might be readily pulverized. Upon microscopic examination, they also exhibited very fine silicious particles, but upon discovering that they could be instantly formed into an unctuous paste, when remixed with water, we have on that account ventured to retain the term.

END OF THE FIRST PART.

DRURY, Tooks Court, Chancery Lane.

PL. 1.

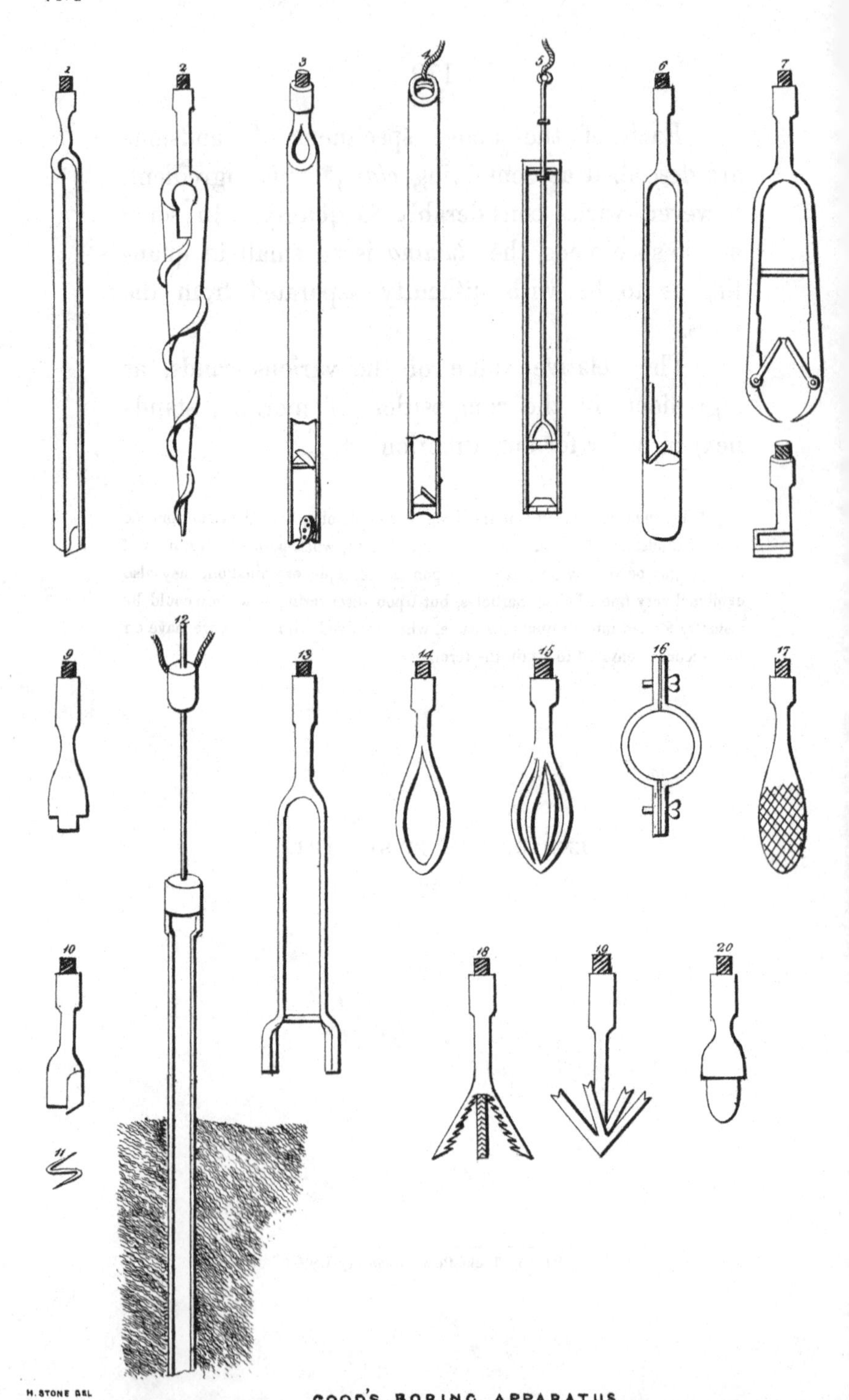

H. STONE DEL.

GOOD'S BORING APPARATUS.

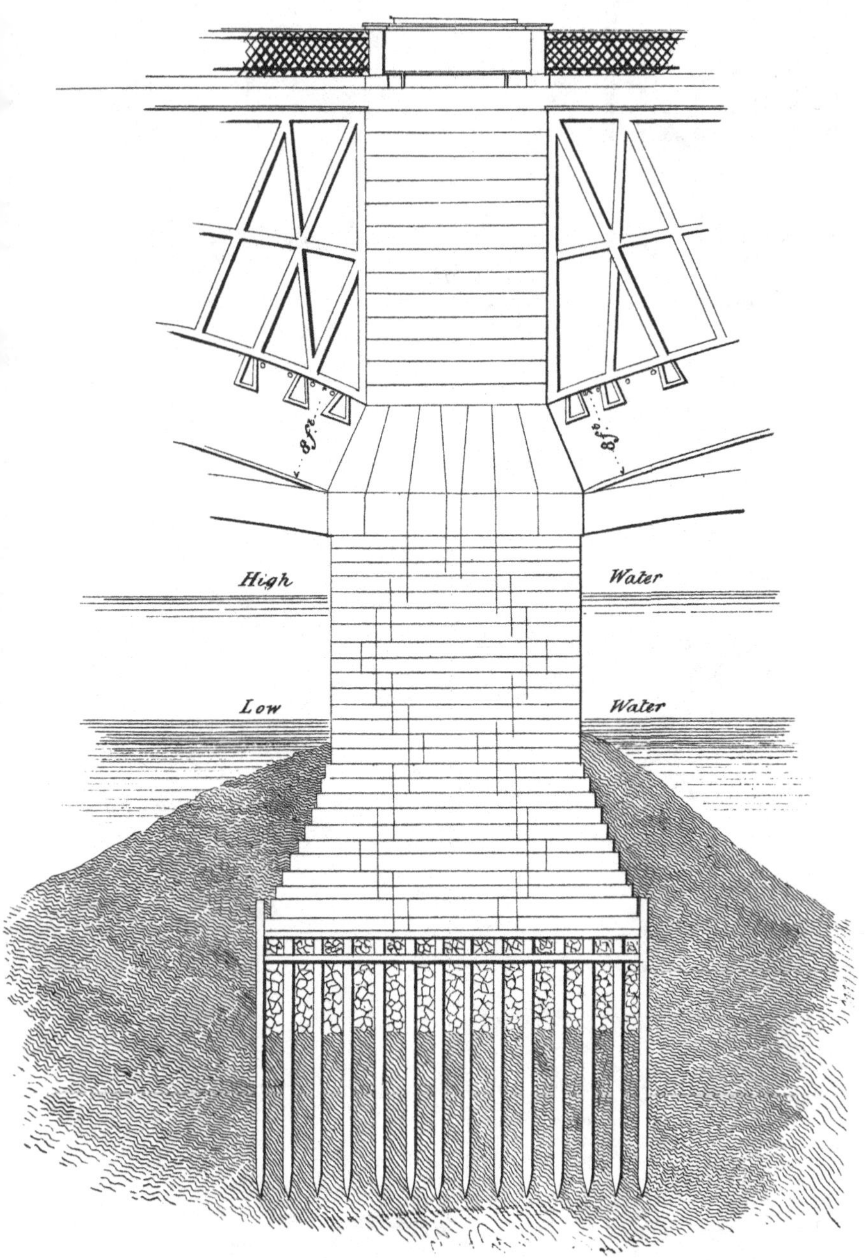

H. STONE DEL.T

SOUTHWARK IRON BRIDGE PIERS.

Pl. 3.

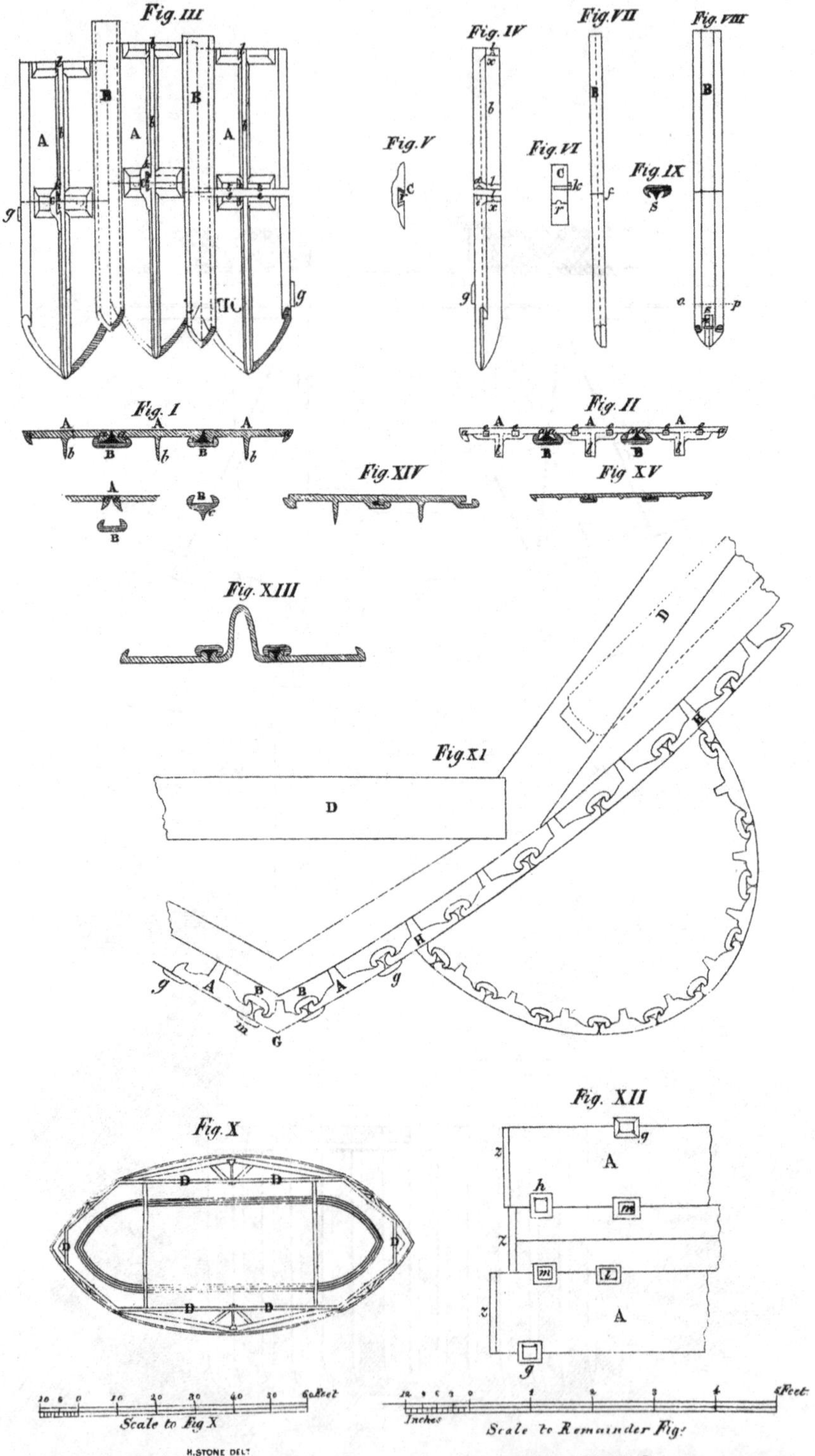

Fig. III

Fig. IV

Fig. V

Fig. VI

Fig. VII

Fig. VIII

Fig. IX

Fig. I

Fig. II

Fig. XIV

Fig. XV

Fig. XIII

Fig. XI

Fig. X

Fig. XII

Scale to Fig X

Inches

Scale to Remainder Figs

H. STONE DELT

IRON PILING.

PL 4

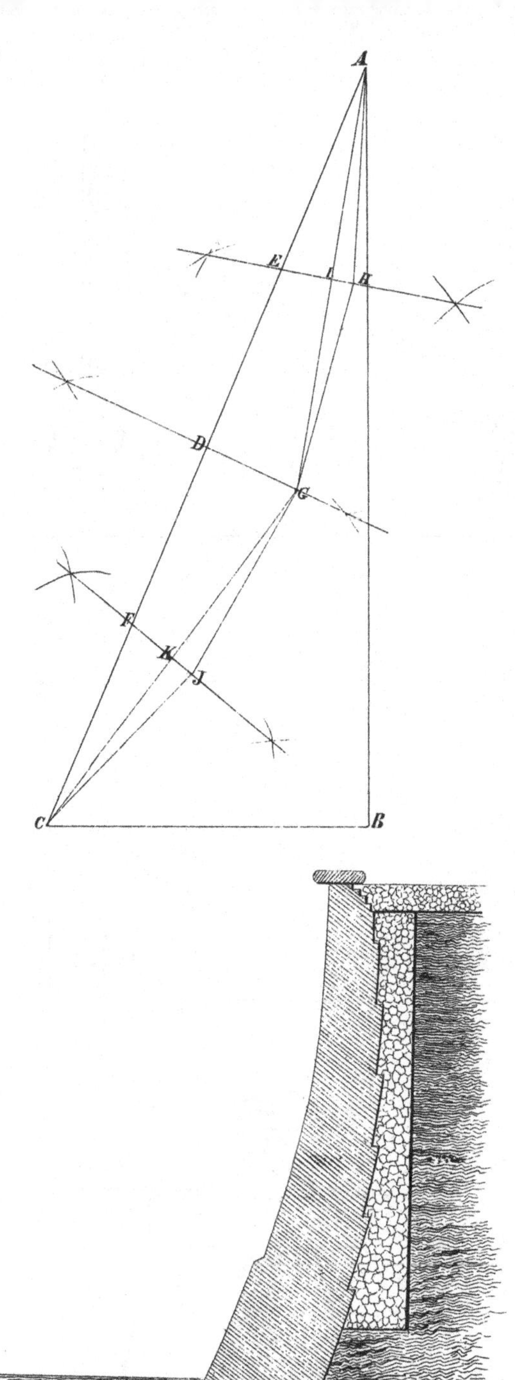

H. STONE. DELT.

CURVATURE OF WINC

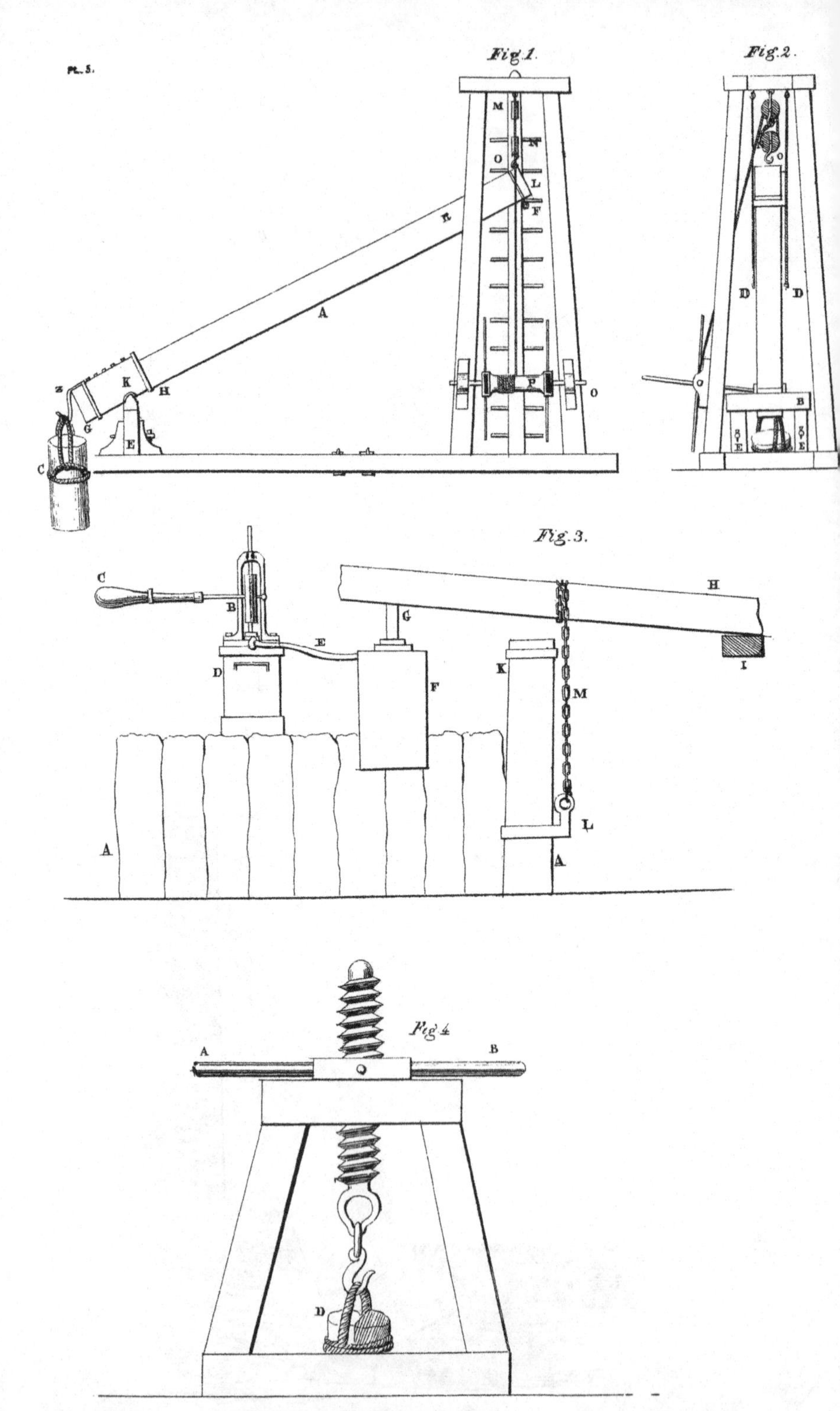

WITHDRAWING MACHINERY

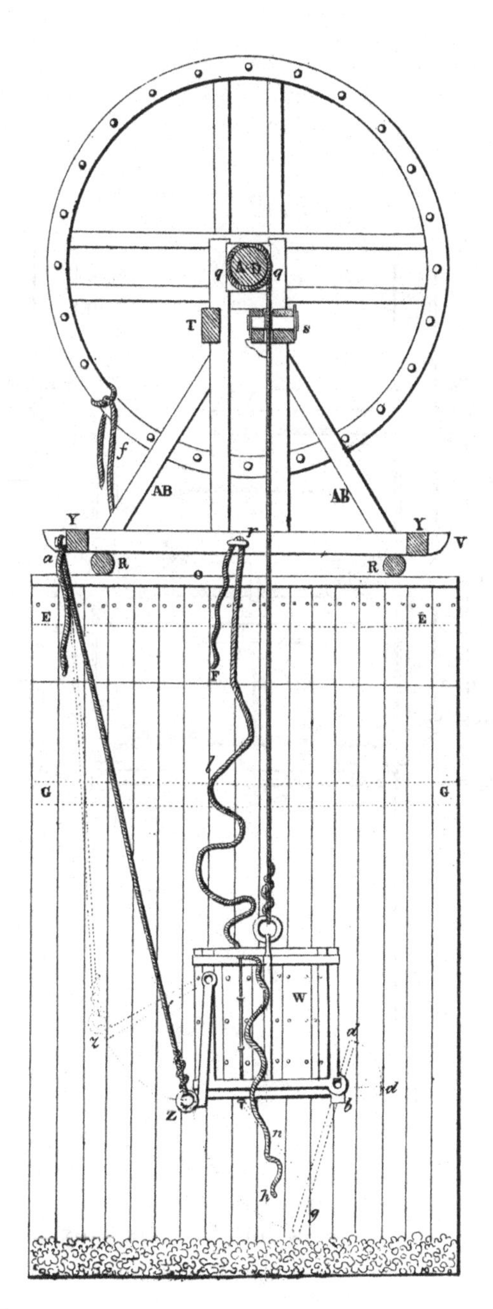

H. STONE, DEL.T

SIDE ELEVATION OF THE MACHINE.

for immersing the Beton"

PL. 7.

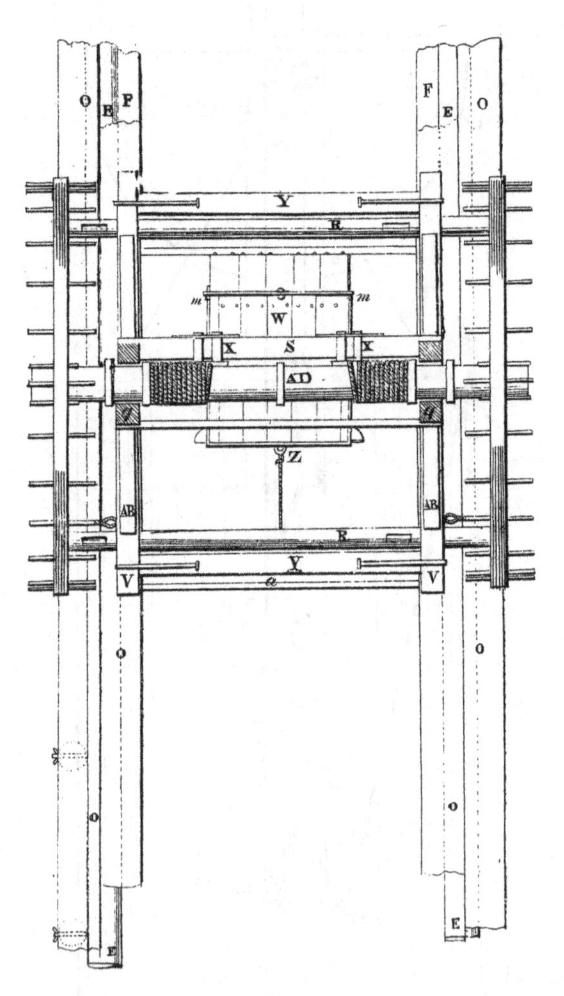

H. STONE DEL^t

PLAN OF THE MACHINE

for immersing the "Beton."

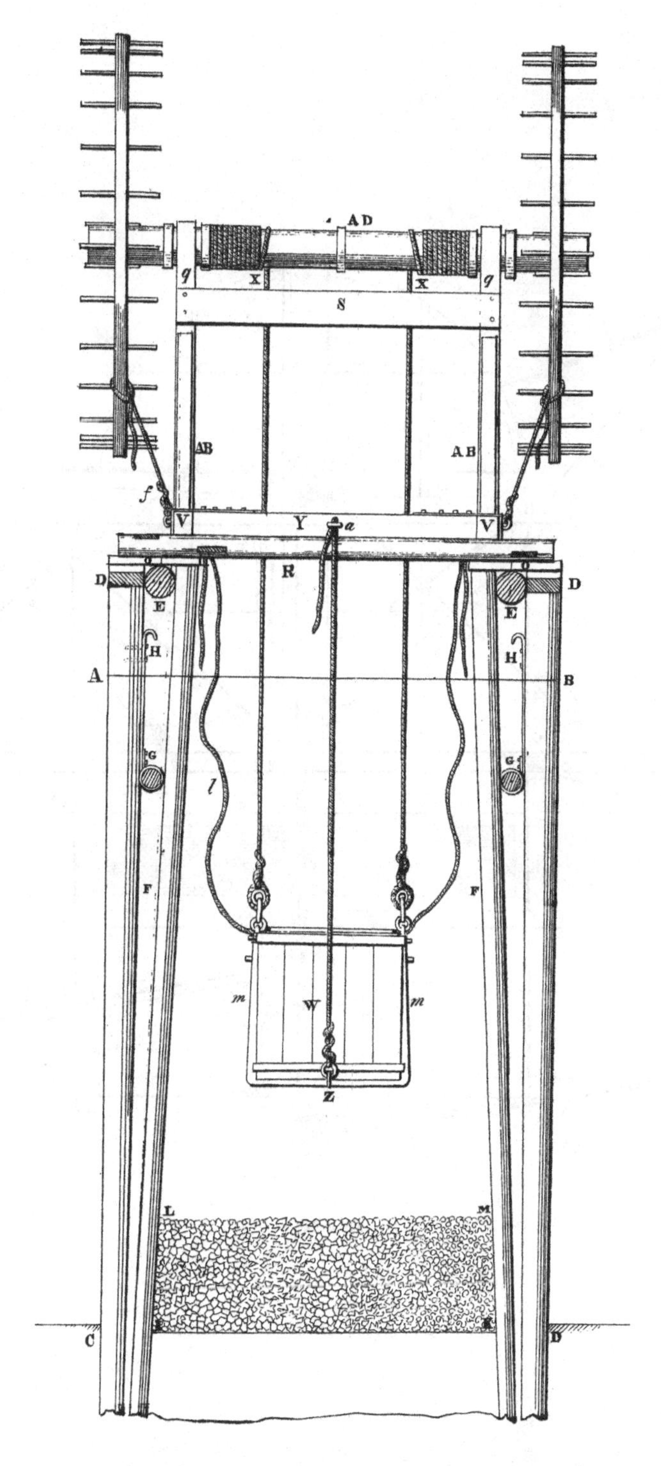

H. STONE DEL.T

LONGITUDINAL ELEVATION OF THE MACHINE.

PL. 9.

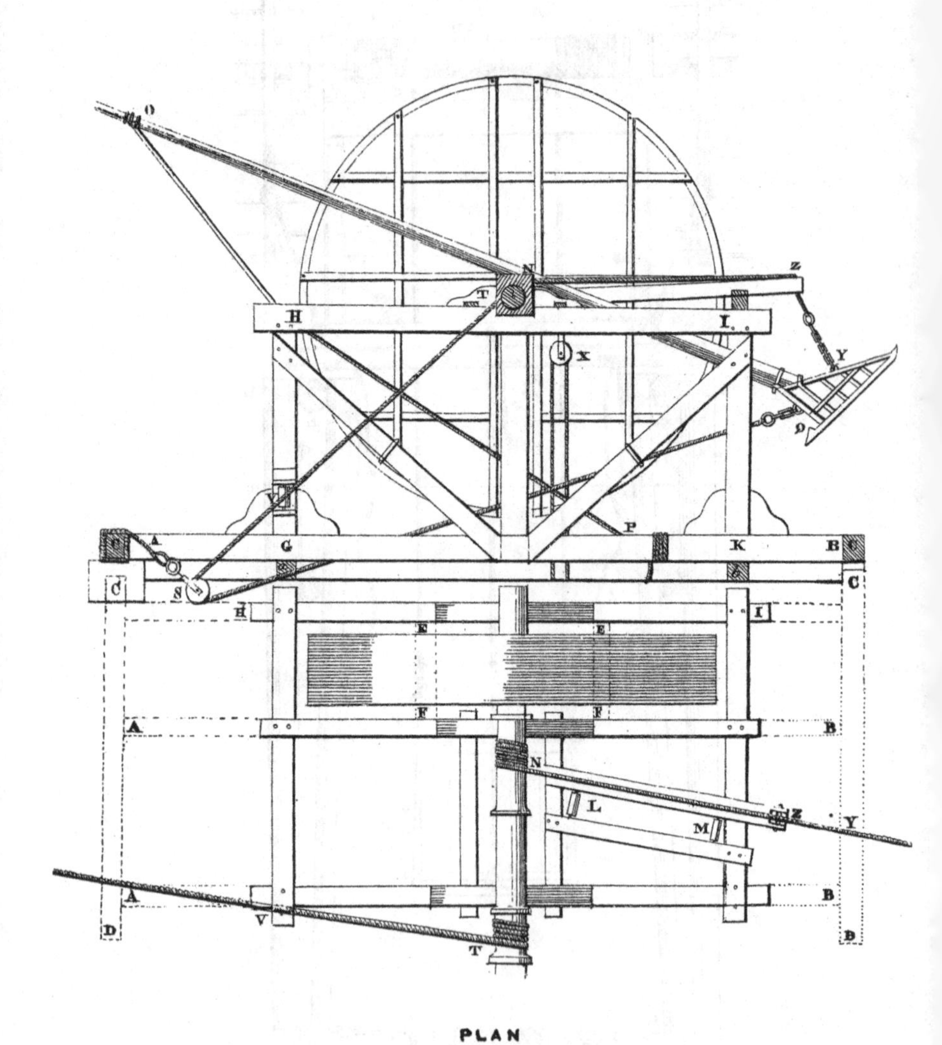

PLAN

H. STONE DEL^T

ELEVATION OF A MACHINE

for cutting Trenches, for Foundations under Water.

HEATHORN'S KILN.

YORKSHIRE KILN.

FIG. 1.

FIG. 3.

SCOTTISH KILN.

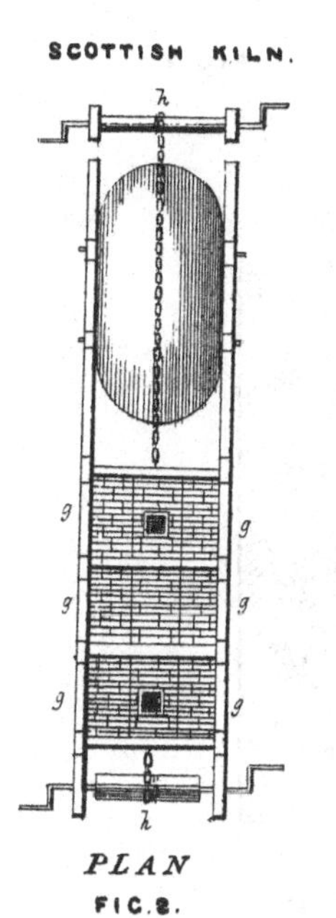

PLAN

FIG. 2.

FIG. 4.

FIG. 5.

H STONE. DEL

KILNS FOR CALCINING LIMESTONE.

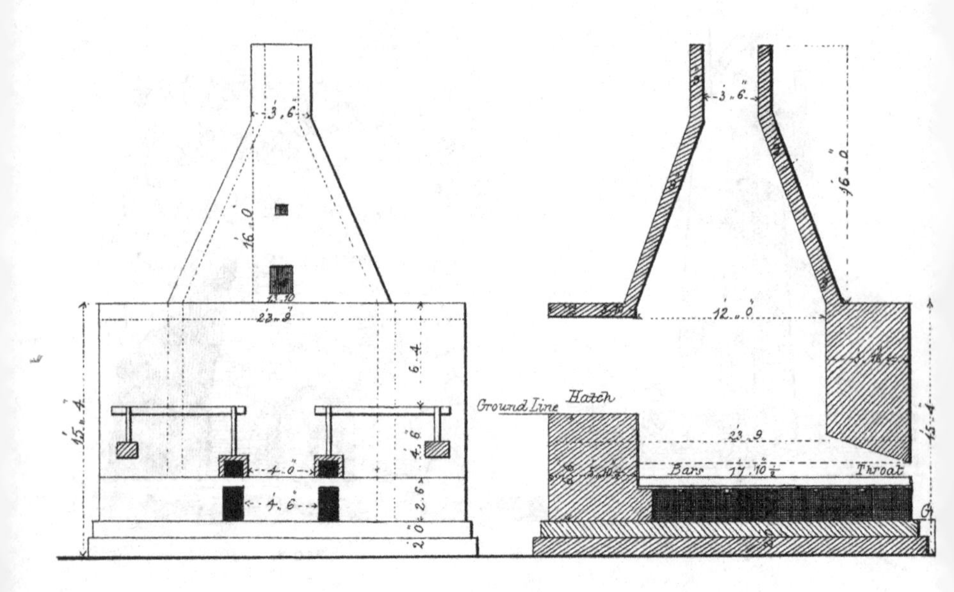

FIG. 1

FIG. 2.

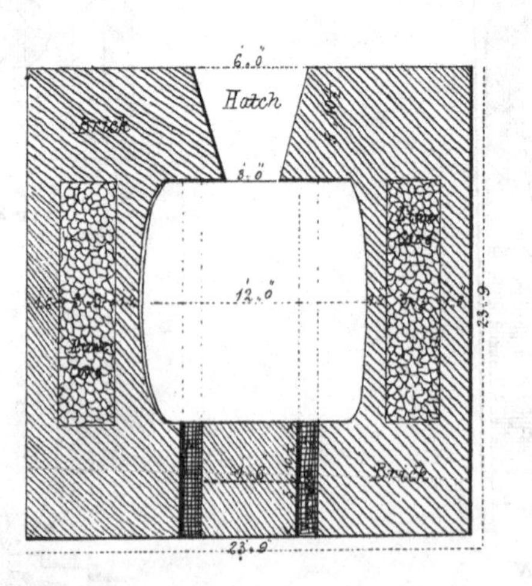

FIG 3.

H. STONE DEL.

DORKING LIME KILN.

COLOUR OF CLAYS

From Sand, Sandstone, Granite &c.

WHEN SEPARATED FROM THE STONE.

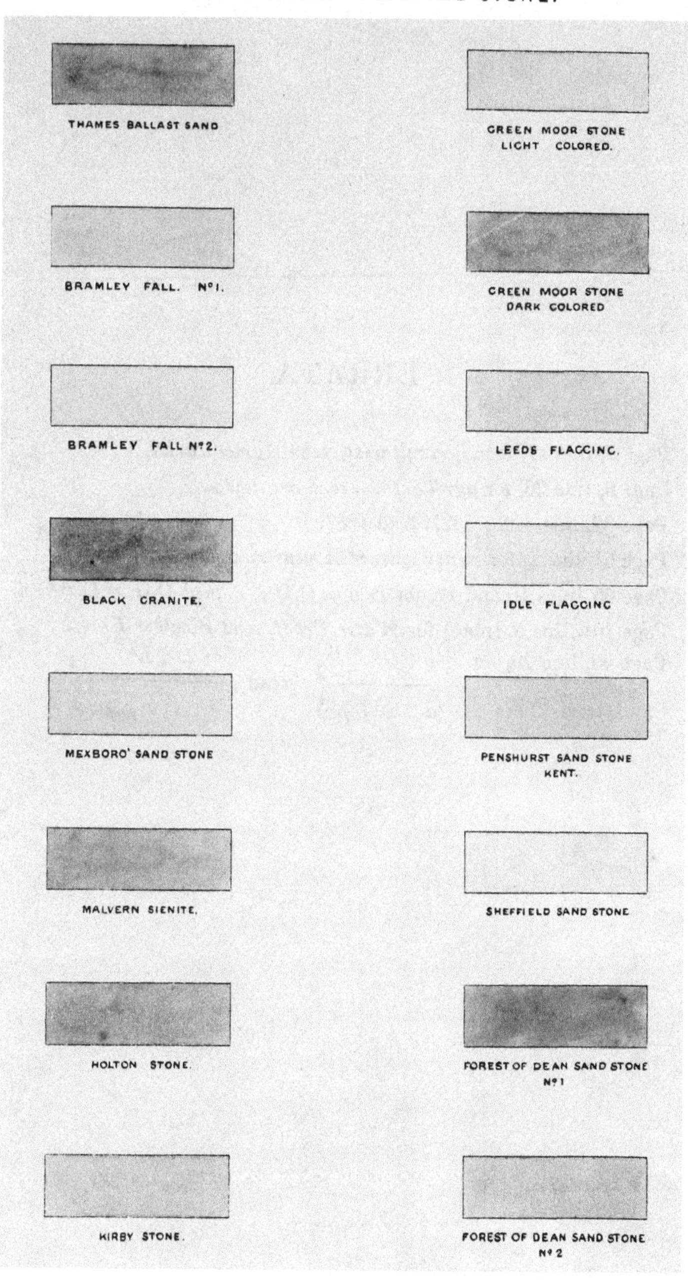

ERRATA.

Page 5, line 17, for *Agricultarist*, read *Agriculturist.*

Page 6, line 26, for *agrillaceous*, read *argillaceous.*

Pεge 32, line 4, for 1827, read 1837.

Page 51, line 1, for *menstrums*, read *menstruums.*

Page 65, lines 16 and 17, for D.G = C.D ÷ 4, read D.G = C.B ÷ 4.

Page 101, line 5, (note) for *Water Proof*, read *Weather Proof.*

Page 49, line 29, for $\left.\dfrac{w \,.\, {}^2h}{{}^2d \,.\, w+p}\right\}$ read $\left.\dfrac{w \,.\, h^2}{d^2\,\overline{w+p}}\right\}=$

Printed in the United States
By Bookmasters